KASURI METHI

Seed Production Technology for Sustainable Horticulture

The Authors

Dr. S.M. Ghawade has been engaged in teaching of various courses to Undergraduate and Post graduate courses, at College of Agriculture and College of Horticulture, Akola, respectively, during 2001 to 2012. He has published more than 55 research papers in Journals of national and international repute, 87 popular articles in different magazines, news papers, contributed 2 chapters in books on Spice crops. He wrote three booklets in English and three in Marathi on various aspects of Horticulture. He had also published twenty seven leaflets on sustainable production technologies of various vegetables with University publication numbers in local language for the immediate benefit to the farmers of the region. He was honored by prestigious K.G. Joshi cash prize award with the hands of Hon'ble M.S. Swaminathan, "Father of green revolution" in India, during 2015. This award was given to him for special efforts of horticultural scientist for the development of production technology in seed spices like Ajwain, Fennel, Coriander and Cumin in western and central zone of Vidarbha. He has guided seventeen post graduate and five Doctoral students in the field of Horticulture and Vegetable science.

Dr. Sanjula Vilas Bhavar (SMS, Horticulture) born at Vedhe, a small village surrounded by agriculture and forest lands in the Palghar district of Maharashtra. Completed his higher secondary education from the A.L.C.College of Arts and Science, Wada, Palghar district of Maharashtra. His parents belongs to farming background motivated him to pursue undergraduate degree in Horticulture (2009-13) and master's degree (2013-2015) in the discipline of vegetable science from the renowned Agriculture University Dr. Balasaheb Sawant Kokan Krishi Vidyapeeth Dapoli, Dist.Ratnagiri, Maharashtra. His master's research was focused on growth and yield parameters of different okra genotypes. He completed his Ph.D. Horticulture (Vegetable Science) from Dr. Panjabrao Deshmukh Krishi Vidyapeeth Akola, Maharashtra. Ph.D. research was focused on introducing kasuri methi at Maharashtra region. For the Ph.D. degree education, he awarded with National Fellowship for higher education, which is given by Government of India. He has cracked two times ARS-NET exam in the discipline of vegetable science. He has published several research papers, popular articles, delivered TV talks, radio talks related to Vegetable science. He has participated and presented papers in various international and national conference, seminars and science congress on horticulture during the year 2015-2018. He worked on different posts as Junior Research Fellow, Senior Research Assistant and Assistant Professor as contractual basis. Currently he is working as Subject Matter Specialist (SMS) of Horticulture at Krishi Vigyan Kendra Khamgaon. Tal. Georai, Dist.Beed, Maharashtra, under Vasantrao Naik Marathwada Krishi Vidyapeeth Parbhani, Maharashtra.

KASURI METHI
Seed Production Technology for Sustainable Horticulture

S.M. Ghawade
M.Sc. (Agriculture) Ph.D. (Horticulture)
Professor of Horticulture, Chilli & Vegetable Research Unit,
Dr. Panjabrao Deshmukh Krishi Vidyapeeth
Akola -444104 (M.S.)

Sanjula Vilas Bhavar
M.Sc. Ph.D. (Horticulture)
Subject Matter Specialist (Horticulture)
Krishi Vigyan Kendra, Gewrai
Ta-Distt.-Beed (M.S.)-431127

2022
Scholars World
A Division of
Astral International Pvt. Ltd.
New Delhi – 110 002

ISBN: 9789354614910

Published by : **Scholars World**
A Division of
Astral International Pvt. Ltd.
– ISO 9001:2015 Certified Company –
4736/23, Ansari Road, Darya Ganj,
New Delhi-110 002
Ph. 011-4354 9197, 2327 8134
E-mail: info@astralint.com
Website: www.astralint.com

Preface

India is the second largest vegetable producer in the world. Vegetables are grown on hardly 2.5 percent of the total area under cultivation of our country existing productivity of vegetables can't meet the national requirements as on today. The demand of vegetables is increasing day by day as majority of us are vegetarian. Vegetables make diet palatable, nutritive and balanced. They are rich sources of vitamins, proteins, minerals, carbohydrates and roughages. The significance of spices, aromatic and medicinal plants have been well known since time immemorial in India.

The seed spices have emerged as one of the important group of spices crops in India. India is the largest producer, consumer and exporter of seed spices in the world. The seed spices accounts for about 36 % and 17 % of the total area and production of spices in the country.

Kasuri methi (*Trigonella Corniculata* L.) is one of the important seed spices crop, slightly differ than regular methi (*Trigonellafoenum – graecium* L.) mostly cultivated for its significant flavor and other medicinal properties in its leaves. It is mostly cultivated in Rajasthan, Gujarat, Bihar, Uttar Pradesh in India.

Though every vegetarian and non vegetarian dishes in India knows importance of Kasuri methi for its magnificent taste and aroma, most of the population doesn't know the difference between regular methi and its procedure of preparations. Now a day, very big Indian companies like MDH, Tata, Everestetc enter into the processing and marketing of Kasuri methi, not only in Maharashtra but throughout the country and export too.

Production technology of Kasuri Methi and its conversion into fragrant, tasty processed leaves is comparatively very simple. The reason identified for less area under cultivation of Kasuri methi in state like Maharashtra, distribution chain from producer to consumer through marketing were taken into consideration. In this regards, genesis for the present study was to give the production technology of seed production of kasuri methi to the farmers of the region. Once they get quality seed material, leaves production and its processing would be acquainted for its marketing chain not only in Vidarbha and state of Maharashtra but, in the country too. We are sure that, this publication shall prove highly useful to various stake holders, such as field functionaries, growers, students and extension workers and those who are having interest in grown Kasuri methi.

We hope that, this book provides technical information for the seed production of Kasuri methi, which would be helpful for others to use quality seed towards improved leaves production and its processing techniques so as to earn more income from small area as a step forward to doubling the farmers income in coming era.

Dr. S. V. Bhavar
Dr. S. M. Ghawade

CONTENTS

CHAPTER 1

INTRODUCTION

Fenugreek is one of the oldest cultivated spice crop of the world and grown for its medicinal value and forage in India, western Asia and Nile valley since remote antiquity. It's a part of Indian diet for over 3000 years. The genus *Trigonella* is reported to have 50 species (more of them originated in the Iranian, Indian region and Oriental origin), among these, eleven species is reported to occurs in India.

1.1 BACKGROUND INFORMATION

The genus *Trigonella* (Family-Leguminaceae) consists of two economically important species, *viz. Trigonella foenum graecum* L. (Common methi) and *Trigonella corniculata* L. (Kasuri methi). Kasuri methi is also called as 'champa methi', 'marwari methi' (Anupama, 2012), 'nagauri methi', 'paan methi' or 'scented methi' (Godara *et al.* 2013). It is originated in Mediterranean region and near East countries. Being a leguminous vegetable, it fixes atmospheric nitrogen in the soil which helps in improving soil fertility.

Trigonella corniculata L. is annual spice crop growing to the height of 60 cm (2ft). Leaves are pinnate, leaflets 1.25-2.0 cm, obovate-cuneate, flowers are hermaphrodite, yellow in colour having close racemes type inflorescence. Pods are 1.2-2.0 cm long, sickle shaped, 4 to 8 seeded. Kasuri methi has a diploid number of chromosomes 2n=16 (Anon., 2004).

It is a semi-arid crop, mainly grown as a *rabi* season crop and the cultivation methods of kasuri methi is more or less similar to that of common methi. It is an herbaceous, bushy, slow growing annual spice crop. It is suitable for light sandy, medium loamy and heavy clay soils and grown well in soil having pH neutral to alkaline. It cannot grow in the shade and remains rosette condition during most of the vegetative growth period. It prefers dry or moist soil (Anon., 2015).

The crop is mainly grown in India, Pakistan, China, Nepal and Bangladesh But, India is the largest producer in the world. In India, it is mainly grown in northern states like Rajasthan, Gujarat, Uttar pradesh, Bihar, Madhya pradesh, Haryana and Punjab. Rajasthan occupies 80 per cent area and production of kasuri methi in the country (Anupama, 2012).

Most renowned Pakistani fenugreek with remarkable aroma cultivated in the 'Kasur' district of Punjab province was known as 'Kasuri methi'. Methi (fenugreek) from 'Kasur' is very famous for its fragrance throughout the country and became geographical indicator of 'Kasur' as 'Kasuri methi' (Erum Shazia *et al.* 2011).

Kasuri methi is a multipurpose crop. Its every part is useful and it utilized in one or other forms as food, fodder, medicine and cosmetics. The dried leaves are economical part of the plant. Dried leaves are used as a spice to add aroma and flavor to the food products. Its seed and tender pods are used as spice for pickle preparation. These are also used in indigenous medicines like diuretic, tonic, carminative, astringent, aphrodisinic (Sharma, 2006). It is gaining importance in recent years and there is a good demand for the dried leaves in the market. The dried leaves are regularly sold in the super markets as popular flavoring material.

Kasuri methi is recognized as a vital source of essential minerals, vitamins and dietary fibers. The green leaves contain several alkaloids like *trigonelline, choline, gentianine* and *carpain.* It contains moisture (86.1 %), protein (4.4 %), fat (0.9 %), fibre (1.1 %), other carbohydrates (6.0 %) and ash (1.5 %) (Anupama, 2012). It also consist of steroidal compound like trigonelline, diosgenin and malonic acid. The bitter test is due to presence of alkaloids, such as trigonelline and choline. Diosgenin has been identified as the major steroidal sapogenin in the leaves of kasuri methi (Varshey and Sood, 1971).

The seeds are rich in proteins, minerals especially Iron, Calcium, Vitamins A, B, C and contains volatile oil, fixed oil, cellulose, starch sugars, alkaloids and enzymes. The volatile oil content in the seed is low (0.02 %), but, it is extremely odoriferous due to a compound called 3-hydroxy-4, 5-dimethyl- oxolane-2 furanone. The fixed oil (7.6 %) has a very bitter taste (Patrick and Yves 1996).

The seeds are carminative, tonic and aphrodisiac and are customarily used for the treatment of colic flatulence, diarrhea, dyspepsia with loss of appetite, chronic cough, dropsy, enlargement of liver and spleen, rickets, gout and diabetes and are good for the elimination of bad breath and bad odor. Infusion of seeds is given to small pox patients as a cooling drink. Infusion of the roasted seeds is used in dysentery. It is also used in sweets served to ladies during post natal period.

In recent past, its high price (100 g dried kasuri methi leaves for 100 rupees) in the market has prompted growers to include this crop in their traditional cropping systems. Also, the importance of fenugreek as seed spice is increasing day by day as its seed are exported in the world market (Rana *et al.* 2015).

1.2 IMPORTANCE OF STUDY

Kasuri methi is not grown commercially in Maharashtra and in Vidarbha region as specific area. But there is commercial market for processed products of kasuri methi which might become profitable business for the farmers and traders. Most of the malls in big cities and at local cities, selling of dried leaves of kasuri methi has been increased day to day. There would be definite demand for this venture in coming future, due to its possibility of production in the Vidarbha region, through group farming concept of State Agriculture Department. Taking into consideration of above facts, cultivation of kasuri methi in the agro-climatic conditions of Vidarbha might become boon to the farmers.

Considering the appropriate agronomical and climatic conditions of Vidarbha region, kasuri methi production might be established with the support of farmer and entrepreneurs. With the increasing area under kasuri methi, processing and final produce should be established by the institute and entrepreneur, so that demand might be established in the local market. Finally, with the increasing area under kasuri methi might have a definite importance in increasing the economic and social status of farmers of the region.

Already, some of the renowned companies viz. Everest, MDH, Badshah, A-1, Suhana, TATA Sampann, National etc. have established there product of kasuri methi in the market. Out of these, any single company if come forward for giving the linkage in production, processing and marketing through the farmers of the region, it would have the potential to change the scenario of production of this crop in the region.

With the support of above fact, there would be the definite importance to fulfil the demand of seeds of kasuri methi. Very limited scientific research were undertaken by Agriculture Universities or Institute in the country on this crop so far. However, sufficient literature is available on sowing dates and fertilizers management in case of regular fenugreek (*Trigonella foenum graecum* L.). Considering this background and importance of seed production of kasuri methi, effort are being undertaken with this experimentations to find out suitable sowing time and appropriate fertilizer dose of primary nutrients in kasuri methi entitled "Effect of sowing time and fertilizer levels on seed yield of kasuri methi (*Trigonella corniculata* L.)".

Hence, the various approaches related to the seed production of kasuri methi i.e. time of sowing and fertilizer doses needs to confirm by conducting scientific experiment.

1.3 OBJECTIVES OF STUDY

The present study entitled "Effect of sowing time and fertilizer levels on seed yield of kasuri methi (*Trigonella corniculata* L.)" was undertaken to assess the efficiency of sowing time and fertilizer dose for the agro-climatic conditions of Vidarbha region with the following objectives.

1. To study the different sowing time and fertilizer levels for seed production of kasuri methi.
2. To find out best treatment combination of sowing time and fertilizer level for seed production of kasuri methi.

1.4 HYPOTHESIS

Considering its magical fragrance and ability to change the taste of vegetarian as well as non-vegetarian dishes, kasuri methi would become profitable business for the farmers and traders. Due to such immense potential, a cultivation of kasuri methi in the agro-climatic conditions of Vidarbha might be the more beneficial and it would be additional cash crop in the existing list of vegetables for the farmers.

Kasuri methi needs cool climatic conditions for its vegetative growth initially. After reproductive stage i.e. flower and pod formation, the quality seed production of this crop would favours moderately hot and dry climatic conditions. Fortunately, Vidarbha region of Maharashtra has mild winter during October and some hot and humid conditions in December, which become hot and dry after mid-January, up to last week of June. Hence, sowing dates related to times of sowing of kasuri methi needs to finalise towards its suitability for adequate vegetative and reproductive growth.

Fertilizer management of leguminous crop has its own requirement of phosphatic and potasic fertilizers as compared to nitrogenous fertilizers. Being a leguminous crop, kasuri methi seed production would be in the line of requirement of fertilization for regular fenugreek. Hence, qualitative seed production of kasuri methi might be responsive for change in levels of primary nutrients as against common methi.

1.5 SCOPE AND LIMITATION

Seed production of spices and condiments are having their specific importance in export and import business of the country. Considering the nutraceutical and medicinal values of seeds of kasuri methi, its qualitative seed production would definitely scrutinized for its production technology especially time of sowing and nutrient management. Considering this fact, till now scientific information, technical knowledge in respect of the quality seed production of kasuri methi in Maharashtra and Vidarbha is badly lacking

There were certain exceptions; i.e. release of some varieties in kasuri methi from State Agriculture Universities and Central Institute of Northern part of the country. Also they have published some scientific literature on this commodity.

But only production of huge quantity of leaves and seed of kasuri methi would serve the purpose of economical strengthening of farmers of this region. Until and unless, leaves of kasuri methi doesn't get flow sheet of proper processing into dried leaves without disturbing its natural colour and aroma; it's further expansion in the fields of farmer would be restricted.

Hence, considering the above limitations the existing agro-climatic conditions of Vidarbha region, well-educated farmers group as well as established seed spices market and processing industries in Maharashtra should get linked with each other, which might have a definite scope in increasing seed and leaves production of kasuri methi in the near future.

Therefore, in context to above, kasuri methi leaves and seed production has a definite scope in Vidarbha region, through such scientific experiments of which sowing times and fertilizers are helpful for increasing seed yield and lucrative return.

2

CHAPTER

REVIEW OF LITERATURE

Sowing time and plant nutrition are the major factors that influence the growth, development and yield of the crops. Kasuri methi (*Trigonella corniculata* L.) is one of the herbaceous, bushy, slow growing annual spice crop mainly grown for dry herb. Dried leaves are used as a spice to add aroma and flavour to the food products. Seeds are used as condiments in several vegetables as well as in pickle preparation (Gurjar, 2005) while, it is also used for medicinal purpose (Sharma *et al*. 2006). The available information regards to plant nutrition and sowing time on kasuri methi is very meager.

Hence, literature related to seed production of kasuri methi with special emphasis to sowing time and nutrition from the available literature related to species, *Trigonella foenum- graecum* L. and other vegetable crops were reviewed under appropriate headings and sub-headings.

- **2.1 Relation of vegetative growth with the dates of sowing and fertilizer doses in kasuri methi**
  - **2.1.1 Effect of dates of sowing on plant height**
  - **2.1.2 Effect of dates of sowing on branches per plant**
  - **2.1.3 Effect of dates of sowing on leaves per plant**
  - **2.1.4 Effect of fertilizer doses on plant height**
  - **2.1.5 Effect of fertilizer doses on branches per plant**
  - **2.1.6 Effect of fertilizer doses on leaves per plant**
  - **2.1.7 Interaction effect of dates of sowing and fertilizer doses on plant height**
  - **2.1.8 Interaction effect of dates of sowing and fertilizer doses on branches per plant**

- 2.1.9 Interaction effect of dates of sowing and fertilizer doses on leaves per plant
- **2.2 Relation of reproductive growth with the dates of sowing and fertilizer doses in kasuri methi**
  - 2.2.1 Effect of dates of sowing on flowering
  - 2.2.2 Effect of dates of sowing on pods per floret, seeds per pod and florets per plant
  - 2.2.3 Effect of fertilizer doses on flowering
  - 2.2.4 Effect of fertilizer doses on pods per floret, seeds per pod and florets per plant
  - 2.2.5 Interaction effect of dates of sowing and fertilizer doses on flowering
  - 2.2.6 Interaction effect of dates of sowing and fertilizer doses on pods per florets, seeds per pod and florets per plant
- **2.3 Relation of seed yield with the dates of sowing and fertilizer doses in kasuri methi**
  - 2.3.1 Effect of dates of sowing on seed yield in kasuri methi
  - 2.3.2 Effect of fertilizer doses on seed yield in kasuri methi
  - 2.3.3 Interaction effect of dates of sowing and fertilizer doses on seed yield in kasuri methi
- **2.4 Relation of seeds and leaves quality with the dates of sowing and fertilizer doses in kasuri methi**
  - 2.4.1 Effect of dates of sowing on seeds and leaves quality
  - 2.4.2 Effect of fertilizer doses on seeds and leaves quality
  - 2.4.3 Interaction effect of dates of sowing and fertilizer doses on seeds and leaves quality in kasuri methi
- **2.5 Relation of soil and plant nutrient status with the dates of sowing and fertilizer doses in kasuri methi**
- **2.6 Economic relation of dates of sowing and fertilizer doses on seed production in kasuri methi**

2.1 RELATION OF VEGETATIVE GROWTH WITH THE DATES OF SOWING AND FERTILIZER DOSES IN KASURI METHI

2.1.1 Effect of dates of sowing on plant height

Bhardwaj (1996) reported that, the maximum (41.40 cm) plant height in french bean (*Phaseolus vulgaris* L.) cultivars, Contender and Pant Anupama was attained, when crop was sown on 20th June, under Nauni (Solan) conditions.

Under Solan (H. P.) conditions, in an experiment conducted on fenugreek (*Trigonella foenum-graecum* L.), Korla and Saini (2003) mentioned that, the crop sown on 6th October had produced the maximum plant height as compared to other dates of sowing under the study.

In the same year 2003, Lal *et al* reported that, the maximum plant height was produced by crop sown on 31st October fenugreek variety MH-23 as compare to 16th November sowing date, under Hissar (Haryana) conditions.

Meena *et al.* (2006) noted that, the height of coriander (*Coriandrum sutivum* L.) increased significantly with an early sowing i.e. 15th October as compared to the delayed sowing, under Udaipur (Rajasthan) conditions.

In an experiment conducted for two years, Nandre *et al.* (2011) reported that, when the fenugreek crop sown on 1st November, gave the maximum 80.00 cm and 82.13 cm plant height in first and second years of experimentations, respectively, under Akola (M. S.) conditions.

Anupama (2012) found that, significantly the maximum plant height at 30, 45 and 60 DAS (7.54 cm, 16.29 cm and 17.57 cm, respectively) were attained, when crop sown on 28th November 2011, with the spacing of 30 cm x 20 cm in kasuri methi (*Trigonella corniculata* L.), under Arabhavi (Karnataka) conditions.

On the basis of two years experimentation (2012 and 2013) in fenugreek, under Egypt conditions, Abou-Shleel (2014) reported that, the maximum plant height in the earlier sowing date i.e. 1st October during both the year of experimentations. And further, they reported that, the plant height was reduced due to delay in sowing of the crop.

Under Peshawar (Pakistan) conditions, in an experiment conducted on coriander (*Coriandrum sutivum* L.), Ali *et al.* (2015) registered that, the sowing done on 5th November had produced the maximum (118.41 cm) plant height as compared to other dates of sowing under the study.

Sowmya (2016) conducted an experiment on effect of sowing dates and stage of pinching on growth, yield and quality of fenugreek and reported that, significantly the maximum plant height at 45, 75 and at last harvest (25.31 cm, 55.26 cm and 69.35 cm, respectively), when crop sown on 10th October were obtained in fenugreek, under Mandasaur (M. P.) conditions.

Meena *et al.* (2018) conducted the experiment on influence of different dates of sowing, fertilizer levels and weedicides on growth and seed yield of fenugreek under semi-arid conditions and reported that, significantly the maximum plant height at 60, 90 and at last harvest (32.61 cm, 61.48 cm and 70.10 cm, respectively) was obtained, when the crop sown on 30th October, under Ajmer (Rajasthan) conditions.

2.1.2 Effect of dates of sowing on branches per plant

Pan *et al.* (2002) recorded that, the maximum branches plant^{-1} of coriander (*Coriandrum sutivum* L.) were produced with 5th November sown crop as compared to other dates of sowing, under Bangalore conditions.

Korla and Saini (2003) recorded that, the sowing of fenugreek (*Trigonella foenum-graecum* L.) done on 6th October had produced significantly the maximum branches plant^{-1} as compared to other dates of sowing, under Solan (H. P.) conditions.

Meena *et al.* (2006) noted that, the number of branches plant^{-1} of coriander increased significantly with an early sowing i.e. 15th October as Compared to the delayed sowing, under Udaipur (Rajasthan) conditions.

Nandre *et al.* (2011) conducted an experiment on effect of sowing dates and nutrient management on growth and seed yield in fenugreek, under Akola (M. S.) conditions with cultivar Pusa Early Bunching and reported that, the crop sown on the 1st November gave maximum number of branches plant^{-1} 14.11 and 15.05 in first and second years of experimentations, respectively.

Anupama (2012) found that, significantly the maximum number of branches at 30, 45 and 60 DAS (3.51, 5.56 and 6.86, respectively) were produced when crop sown on 28th November, with the spacing of 30 cm x 20 cm in kasuri methi (*Trigonella corniculata* L.), under Arabhavi (Karnataka) conditions.

On the basis of two years experimentations (2012 and 2013) in fenugreek, under Egypt conditions, Abou-Shleel (2014) reported that, the maximum number of branches in the earlier sowing date i.e. 1st October during both the years of experimentations. And further, they reported the reduction in number of branches due to delay in sowing of the crop.

Sowmya (2016) conducted the experiment on effect of sowing dates and stage of pinching on growth, yield and quality of fenugreek and reported that, significantly the maximum number of branches plant^{-1} at 45, 75 and at last harvest (6.32, 7.60 and 16.53, respectively) were obtained when crop sown on 10th October in fenugreek, under Mandasaur (M. P.) conditions.

Meena *et al.* (2018) reported that, significantly the maximum number of primary branches at 60, 90 and at last harvest (4.28, 4.53 and 4.64, respectively) were obtained, when the crop sown on 30th October 2016 in fenugreek, under Ajmer (Rajasthan) conditions.

2.1.3 Effect of dates of sowing on leaves per plant

Nandre *et al.* (2011) studied the effect of sowing dates and nutrient management on growth and seed yield in fenugreek (*Trigonella foenum-graecum* L.) along with Pusa Early Bunching cultivar, under Akola (M. S.) conditions and observed that, different dates of sowing showed significant effect on growth component. The maximum number of leaves plant^{-1} (44.99 and 46.96) were observed, when the crop sown during 1st November of first and second years of experimentation, respectively.

Anupama (2012) found that, significantly the maximum number of trifoliate leaves at 30, 45 and 60 DAS (12.24, 51.52 and 61.59, respectively) were obtained in kasuri methi (*Trigonella corniculata* L.), when the crop sown on 28th November 2011, with the spacing of 30 cm x 20 cm, under Arabhavi (Karnataka) conditions.

On the basis of two years experimentations (2012 and 2013) in fenugreek, under Egypt conditions, Abou-Shleel (2014) reported that, the maximum number of leaves were produced in the sowing date 1st October during both the years of experimentations. And further, they have reported the reduction in number of leaves due to delay in sowing of the crop.

2.1.4 Effect of fertilizer doses on plant height

Bhardwaj (1996) reported that, the maximum (33.44 cm) plant height was obtained, when fertilizer dose applied @ 54:125:37.5 kg ha^{-1} NPK and which was at par (33.42) with the fertilizer dose @ 45:100:30 kg ha^{-1} NPK in french bean (*Phaseolus vulgaris* L.) cultivars, Contender and Pant Anupama, under Nauni (Solan) conditions.

Datta *et al.* (2005) found that, significantly the maximum (53.84 cm) plant height was observed with an application of nitrogen @ 45 kg ha^{-1} in fenugreek (*Trigonella foenum-graecum* L.), grown under Mohanpur (West Bengal) conditions.

Gurjer (2005) reported that, significantly the maximum (115.0 cm) plant height was observed with an application of FYM @ 10 t ha^{-1} in kasuri methi (*Trigonella corniculata* L.) grown, under Gwalior (M. P.) conditions.

Meena *et al.* (2006) noted that, the plant height of coriander (*Coriandrum sutivum* L.) increased significantly with an increasing nitrogen dose i.e. 75 kg ha^{-1} as compared to the low nitrogen dose, under Udaipur (Rajasthan) conditions.

Kumar and Singh (2007) opined that, an application of 50 kg ha^{-1} phosphorus had produced the maximum (53.9 cm) plant height in fenugreek, under IARI, New Delhi conditions.

Deora *et al.* (2009) reported that, significantly the maximum (24.4 cm) plant height at last harvest of leaves with the application of fertilizer dose @ 20:40:00 kg ha^{-1} NPK, in kasuri methi grown, under Bikaner (Rajasthan) conditions.

Nandre *et al.* (2011) conducted an effect of sowing dates and nutrient management on growth and seed yield in fenugreek and reported that, the fertilizer dose applied @ 37.50:18.75 kg ha^{-1} NP and 10 t ha^{-1} FYM gave the maximum plant height 81.52 cm and 84.11 cm, in first and second years of experimentations, respectively.

Anupama (2012) found that, significantly the maximum plant height at 30, 45 and 60 DAS (8.56 cm, 17.31 cm and 19.02 cm, respectively) was obtained with the nitrogen dose applied @ 80 kg ha^{-1} in kasuri methi grown, under Arabhavi (Karnataka) conditions.

Koyani *et al.* (2012) opined that, the maximum plant height was noted under the treatment receiving 120 kg ha^{-1} nitrogen in combination with 60 kg ha^{-1} phosphorus in fennel (*Foeniculum vulgare* Mill.), under Junagadh (Gujrat) conditions.

Godara *et al.* (2013) reported that, the maximum (11.8 cm) plant height was obtained when phosphorus and sulphur dose applied @ 60 kg ha^{-1} and 45 kg ha^{-1}, respectively, in kasuri methi, under Ajmer (Rajasthan) conditions.

Sunanda *et al.* (2014) conducted the experiment on the influence of integrated nutrient management on growth, yield and quality of kasuri methi and reported that, significantly the maximum plant height at 30, 60, 90 DAS (5.51 cm, 27.26 cm, 42.81 cm, respectively) was noticed when fertilizer dose applied @ 60:25:50 kg ha^{-1} NPK + FYM (7.5 t ha^{-1}) + *Rhizobium* (1.5 t ha^{-1}) + *Azospirillum* (5 kg ha^{-1}) + PSB (5 kg ha^{-1}) in kasuri methi, under hill zone of Karnataka.

In the seed production experiment conducted on coriander, Ali *et al.* (2015) noted that, the significant increase in the height of the plant with the increasing dose of nitrogen. Application of nitrogen @ 60 kg ha^{-1} gave the maximum (132.00 cm) plant height, under Peshawar (Pakistan) conditions.

Kumar *et al.* (2015) studied the response of crop geometry, fertilizer levels and genotype on productivity of fenugreek and reported that, significantly the maximum (72.5 cm) plant height at last harvest was found, when fertilizer dose was applied @ 50:50:25 kg ha^{-1}, under Ajmer (Rajasthan) conditions.

In one year experimentation, Singh *et al.* (2017[b]) noticed that, the plant height was increased significantly due to an application of 60 kg ha^{-1} phosphorus in fenugreek, under Meerut (U. P.) conditions.

On the basis of two years experimentations, Datta *et al.* (2017) recorded that, an application of 80 kg nitrogen, 80 kg phosphorus and 40 kg potash ha^{-1} had recorded the maximum (75.74 cm) plant height at 75 DAS in fenugreek, under Mohanpur (West Bengal) conditions.

Meena *et al.* (2018) reported that, significantly the maximum plant height at 60, 90 and at last harvest (30.74 cm, 61.05 cm and 67.57 cm, respectively) was noticed, when fertilizer dose was given @ 50:50:25 kg ha^{-1} NPK in fenugreek, under Ajmer (Rajasthan) conditions.

Mitoo *et al.* (2018) conducted an experiment on fertilizer dose related to seed production in fenugreek, under Sher-e-Bangla Agricultural University (Bangladesh) and concluded that, application of phosphorus 40 kg ha^{-1} and Sulphur 10 kg ha^{-1} gave the maximum plant height at 30 DAS.

2.1.5 Effect of fertilizer doses on branches per plant

Gurjer (2005) reported that, significantly the maximum (8.00) number of branches with the application of 40:40:20 kg ha^{-1} NPK + FYM 5 t ha^{-1} was observed in, kasuri methi (*Trigonella corniculata* L.), grown under Gwalior (M. P.) conditions.

Meena *et al.* (2006) noted that, the number of branches plant^{-1} in coriander (*Coriandrum sutivum* L.) increased significantly with an increasing nitrogen dose i.e. 75 kg ha^{-1} as compared to the low nitrogen doses, under Udaipur (Rajasthan) conditions.

Kumar and Singh (2007) opined that, an application of 50 kg ha^{-1} phosphorus had given the maximum (18.0) branches plant^{-1} in fenugreek (*Trigonella foenum-graecum* L.), under IARI, New Delhi conditions.

Deora *et al.* (2009) reported that, significantly the maximum (7.9) number of branches plant^{-1} were produced with the application of fertilizer dose @ 20:40:0 kg ha^{-1} NPK, in kasuri methi, grown under Bikaner (Rajasthan) conditions.

In an experiment conducted for two years, Nandre *et al.* (2011) reported that, when fertilizer dose applied @ 37.50:18.75 kg ha^{-1} NP and 10 t ha^{-1} FYM gave the maximum number of branches plant^{-1} (14.00 and 15.10) in first and second years of experimentations, respectively in fenugreek, under Akola (M. S.) condition.

Anupama (2012) reported that, significantly the maximum number of branches at 30, 45 and 60 DAS (3.83, 6.43 and 7.35, respectively) were obtained in kasuri methi, when the nitrogen apply @ 80 kg ha^{-1}, under Arabhavi (Karnataka) conditions.

Koyani *et al.* (2012) recorded that, the branches plant^{-1} in fennel (*Foeniculum vulgare* Mill.) were increased significantly due to an application of 120 kg ha^{-1} nitrogen in combination with 60 kg ha^{-1} phosphorus, under Junagadh (Gujrat) conditions.

Sunanda *et al.* (2014) conducted an experiment on the influence of integrated nutrient management on growth, yield and quality of kasuri methi and reported that, significantly the maximum number of branches plant^{-1} at 60, 90 and at last harvest (14.02, 23.51 and 25.64, respectively) were produced, when fertilizer dose applied @ 60:25:50 kg ha^{-1} NPK + FYM (7.5 t ha^{-1}) + *Rhizobium* (1.5 t ha^{-1}) + *Azospirillum* (5 kg ha^{-1}) + PSB (5 kg ha^{-1}) in kasuri methi, under hill zone of Karnataka.

Kumar *et al.* (2015) studied the response of crop geometry, fertilizer levels and genotype on productivity of fenugreek and reported that, significantly the maximum (6.0) number of branches at last harvest, when fertilizer dose applied @ 50:50:25 kg ha^{-1}, under Ajmer (Rajasthan) conditions.

On the basis of two years experimentations, Datta *et al.* (2017) recorded that, an application of 60:80:40 kg ha^{-1} NPK, had recorded the maximum number of secondary branches plant^{-1} (15.94) in fenugreek, under Mohanpur (West Bengal) conditions.

In one year experimentation of fenugreek seed production, Singh *et al.* (2017[b]) noticed that, the number of branches plant^{-1} were increased significantly due to an application of 60 kg ha^{-1} phosphorus, under Meerut (U. P.) conditions.

Meena *et al.* (2018) conducted the experiment on influence of different dates of sowing, fertilizer levels and weedicides on growth and yield of fenugreek under semi-arid condition and found that, significantly the maximum number of primary branches plant^{-1} at 60, 90 and at last harvest (3.80, 4.22 and 4.33, respectively) were obtained, when fertilizer dose given @ 50:50:25 kg ha^{-1} NPK, under Ajmer (Rajasthan) conditions.

2.1.6 Effect of fertilizer doses on leaves per plant

Nandre *et al.* (2011) reported that, the maximum number of leaves plant^{-1} (45.47 and 47.93) in first and second years of experimentations, respectively in fenugreek (*Trigonella foenum-graecum* L.) seed production were produced, when fertilizer dose applied @ 37.50:18.75 kg ha^{-1} NP and 10 t ha^{-1} FYM, under Akola (M. S.) conditions.

Anupama (2012) reported that, significantly the maximum number of trifoliate leaves at 30, 45 and 60 DAS (14.34, 62.31 and 71.04, respectively) with the application of nitrogen @ 80 kg ha^{-1} were obtained in kasuri methi (*Trigonella corniculata* L.), under Arabhavi (Karnataka) conditions.

Godara *et al.* (2013) reported that, the maximum (7.7) number of leaves were produced, when phosphorus and sulphur applied @ 60 kg ha^{-1} and 45 kg ha^{-1}, respectively in kasuri methi, under Ajmer (Rajasthan) conditions.

Sunanda *et al.* (2014) conducted the experiment on the influence of integrated nutrient management on growth, yield and quality of kasuri methi and reported that, the maximum number of trifoliate leaves per plant at 30, 60 and 90 DAS (9.14, 58.89 and 86.05, respectively) were noticed, when fertilizer dose applied @ 60:25:50 kg ha^{-1} NPK + FYM (7.5 t ha^{-1}) + *Rhizobium* (1.5 t ha^{-1}) + *Azospirillum* (5 kg ha^{-1}) + PSB (5 kg ha^{-1}) in kasuri methi, under hill zone of Karnataka.

Singh *et al.* (2017[b]) noticed that, the number of leaves at last harvesting stage was increased significantly due to an application of 60 kg ha^{-1} phosphorus and observed 125.33 leaves at last harvesting stage, under Meerut (U. P.) conditions in fenugreek.

2.1.7 Interaction effect of dates of sowing and fertilizer doses on plant height

Bhardwaj (1996) studied seed production in french bean (*Phaseolus vulgaris* L.) cultivars and found that, the maximum (42.48 cm) plant height was attained, when fertilizer dose @ 50:125:37.5 kg ha^{-1} NPK and crop sown on the date of 20th June, under Nauni (Solan) conditions.

Nandre *et al.* (2011) studied the effect of sowing dates and nutrient management on growth and seed yield of fenugreek (*Trigonella foenum-graecum* L.), under Akola (M. S.) conditions and interaction effect of different fertilizer doses and sowing dates, in terms of plant height were found to be non-significant, during both the years of experimentations.

Anupama (2012) reported that, significantly the maximum plant height at 30, 45 and 60 DAS (8.65 cm, 17.33 cm and 19.45 cm, respectively) with the nitrogen dose 80 kg ha^{-1} and spacing of 30 cm x 20 cm was observed, when crop sown on the date 28th November, in kasuri methi (*Trigonella corniculata* L.), under Arabhavi (Karnataka) conditions.

Ali *et al.* (2015) pointed that, the plant height was the maximum in coriander (*Coriandrum sutivum* L.) crop, when it was sown on 5th November and fertilized with 60 kg ha^{-1} nitrogen gave the maximum (133.66 cm) plant height, under Peshawar (Pakistan) conditions.

Meena *et al.* (2018) revealed that, when NPK applied to the fenugreek sown on 30th October with the weedicide (Oxadiargyl) proved to have no effect on the plant height, under Ajmer (Rajasthan) conditions.

2.1.8 Interaction effect of dates of sowing and fertilizer doses on branches per plant

Nandre *et al.* (2011) observed that, interaction effect of different fertilizer doses and sowing dates, in terms of number of branches plant^{-1} found to be non-significant during both the years of experimentations 2004-05 and 2005-06 respectively, in fenugreek (*Trigonella foenum-graecum* L.), under Akola (M. S.) conditions.

Anupama (2012) reported that, significantly the maximum number of branches at 30, 45 and 60 DAS (4.39, 6.67 and 7.37, respectively) with the nitrogen dose 80 kg ha^{-1} and spacing of 30 cm x 20 cm were obtained when, crop sown on the date 28th November, in kasuri methi (*Trigonella corniculata* L.) under Arabhavi (Karnataka) conditions.

Meena *et al.* (2018) studied the influence of different date of sowing, fertilizer level and weedicides on growth and yield of fenugreek under semi-arid conditions. They stated that, NPK applied to the crop sown on 30th October with the weedicide (Oxadiargyl), significantly affected on number of primary branches, under Ajmer (Rajasthan) conditions.

2.1.9 Interaction effect of dates of sowing and fertilizer doses on leaves per plant

Nandre *et al.* (2011) found that, the interaction effect of different fertilizer doses and dates of sowing, in terms of leaves plant^{-1} found to be non-significant during both the years of experimentations 2004-05 and 2005-06 respectively, under Akola (M. S.) conditions.

Anupama (2012) reported that, significantly the maximum number of trifoliate leaves at 30, 45, 60 DAS (14.39, 63.45, 71.18, respectively) with the nitrogen dose 80 kg ha^{-1} and spacing of 30 cm x 20 cm were obtained, when the crop sown on the date 28th November 2011, in kasuri methi, under Arabhavi (Karnataka) conditions.

2.2 RELATION OF REPRODUCTIVE GROWTH WITH THE DATES OF SOWING AND FERTILIZER DOSES IN KASURI METHI

2.2.1 Effect of dates of sowing on flowering

Bhardwaj (1996) studied seed production in french bean (*Phaseolus vulgaris* L.) cultivars and found that, significantly the minimum (32.22) days to 50 per cent flowering, were occurred when the crop was sown on 05th July, under Nauni (Solan) conditions.

Sheoran *et al.* (2000) revealed that, minimum (68.5) days were required for flower initiation and days to maturity (130) in fenugreek genotypes, HM-65 and T-8, when the crop was sown on 16th December, under Hisar conditions.

Meena *et al.* (2006) noted that, the days to initiation of flowering and days to 50 per cent flowering in coriander (*Coriandrum sutivum* L.) were decreased significantly with an early sowing i.e. 15th October as compared to the delayed sowing, under Udaipur (Rajasthan) conditions.

Nandre *et al.* (2011) reported that, the minimum days were required for flowering (45.08 and 45.41) in first and second year of experimentation, respectively in fenugreek, under Akola (M. S.) condition, when crop was sown on the date of 1st November.

Meena *et al.*, (2018) reported that, significantly the minimum days to 50 per cent flowering (51) were needed, when crop sown on 30th October, under Ajmer (Rajasthan) conditions.

2.2.2 Effect of dates of sowing on pods per florets, seeds per pod and florets per plant

Bhardwaj (1996) found that, significantly the maximum days to seed harvest (89.22), number of pods plant^{-1} (14.76), pod length (14.06 cm), number of seeds pod^{-1} (4.76) in cultivars, Contender and Pant Anupama were required, when crop sown on 20th June, under Nauni (Solan) conditions.

Yadav *et al.* (2000) reported that, the seeds pod^{-1} were the maximum with the first date of sowing i.e. 31st October and it was decreased significantly with each 20 days delayed in sowing time in fenugreek (*Trigonella foenum-gracecum* L.).

In the subsequent year Gill *et al.* (2001) reported that, the pods plant^{-1} in fenugreek were decreased significantly with delayed sowing. And significantly the maximum pods plant^{-1} were produced by the crop when it was sown on 1st October, under Ludhiana conditions.

Nandre *et al.* (2011) quoted that, the pods plant^{-1} and seeds pod^{-1} were produced maximum (34.82 and 15.14, respectively) in numbers with first date of sowing (1st November) and it was decreased significantly with each 15 days delayed in sowing time in fenugreek, under Akola (M. S.) conditions.

In two year experimentation, Abou-Shleel (2014) reported that, there were significant differences of sowing dates on vegetative growth of fenugreek. The sowing date 1st October was recorded the maximum number of pods plant^{-1} (9.85) and seeds plant^{-1} (14.65) and subsequently, delayed in sowing resulted into reduction in pods plant^{-1} and seeds plant^{-1}, under Egypt conditions.

Singh *et al.* (2017a) found that, significantly the maximum (57.90) number of pods plant^{-1}, pod length (11.72 cm) and seeds pod^{-1} (16.88) were obtained when, the fenugreek crop was sown on 10th October, under Gwalior (M.P) conditions.

Kauser *et al.* (2018) reported that, the maximum (13.80) number of pod plant^{-1}, pod length (10.34 cm) and seeds pod^{-1} (13.60) were produced, when crop was sown on 15th October, in fenugreek, under Mudigere (Karnataka) conditions.

Meena *et al.* (2018) reported that, the maximum (37.23) number of pods plant^{-1}, pod length (9.95 cm) and seeds pod^{-1} (15.98) were produced, when crop was sown on 30th October, in fenugreek, under Ajmer (Rajasthan) conditions.

2.2.3 Effect of fertilizer doses on flowering

Bhardwaj (1996) reported that, the minimum (36.17) days to 50 per cent flowering were needed, when fertilizer dose was applied @ 36:75:22.5 kg ha^{-1} NPK in french bean (*Phaseolus vulgaris* L.), under Nauni (Solan) conditions.

Sheoran *et al.* (2000) recorded that, minimum (69.5) days were required for flower initiation and days to maturity (138.5) in fenugreek genotypes, HM-65 and T-8, when the phosphorus was applied @ 90 kg ha^{-1}, under Hisar conditions.

Meena *et al.* (2006) noted that, the days to initiation of flowering and days to 50 per cent flowering in coriander (*Coriandrum sutivum* L.) decreased significantly with the decreasing dose of nitrogen i.e. 60 kg ha^{-1} as compare to higher nitrogen dose, under Udaipur (Rajasthan) conditions.

Nandre *et al.* (2011) observed that, different fertilizer doses showed significant effect on reproductive stage. The minimum days to flowering (42.70 and 42.73, respectively) were required in control treatment (no fertilizer applied), during both the years of experimentations.

Sunanda *et al.* (2014) reported that, the minimum (46.25) days to flower initiation and 50 per cent flowering (71.06) were needed when, the fertilizer dose was applied @ 80:25:50 kg ha^{-1} NPK + vermicompost (4 t ha^{-1}). While, the maximum days to flower initiation (56.34) and 50 % flowering (83.72) were taken by kasuri methi crop when, the fertilizer dose was applied @ 60:25:50 kg ha^{-1} NPK + FYM (7.5 t ha^{-1}) + *Rhizobium* (1.5 t ha^{-1}) + *Azospirillum* (5 kg ha^{-1}) + PSB (5 kg ha^{-1}), under hill zone of Karnataka.

On the basis of two years experimentations, Datta *et al.* (2017) recorded that, an application of 40 kg nitrogen, 60 kg phosphorus and 20 kg potash ha^{-1} had recorded the minimum (49.36) days to 50 per cent flowering in fenugreek, under Mohanpur (West Bengal) conditions.

In the same year Singh *et al.* (2017^{b}) noticed that, the minimum (95.66) days were taken to flowering in control treatment (no fertilizer), under Meerut (U. P.) conditions.

Meena *et al.* (2018) reported that, significantly the minimum 53.77) days to 50 per cent flowering were required, when fertilizer dose was applied @ 50:50:25 kg ha^{-1} NPK, in fenugreek, under Ajmer (Rajasthan) conditions.

2.2.4 Effect of fertilizer doses on pods per floret, seeds per pod and florets per plant

Bhardwaj (1996) found that, significantly the maximum (89.63) days to seed harvest, number of pods plant^{-1} (12.83), pod length (12.68 cm), seeds pod^{-1} (4.14) were needed in french bean, when fertilizer dose was applied @ 54:125:37.5 kg ha^{-1} NPK, under Nauni (Solan) conditions.

Datta *et al.* (2005) found that, significantly the maximum (37.63) pods plant^{-1}, pod length (12.60 cm) and seeds plant^{-1} (13.74) were obtained with the application of nitrogen @ 45 kg ha^{-1} in fenugreek (*Trigonella foenum-graecum* L.), grown under Mohnpur (West Bengal) conditions.

Kumar and Singh (2007) opined that, an application of 50 kg ha^{-1} phosphorus had gave the maximum (17.2) pods plant^{-1} and seeds plant^{-1} (16.3) in fenugreek, under IARI, New Delhi conditions.

Nandre *et al.* (2011) recorded that, the maximum pods plant^{-1} (41.60) and seeds pod^{-1} (16.56) were harvested from the fenugreek crop, where application of fertilizer dose @ 37.50:18.75 kg ha^{-1} NP and 10 t ha^{-1} FYM in fenugreek, under Akola (M. S.) conditions.

Koyani *et al.* (2012) recorded that, the umbels plant^{-1} of fennel (*Foeniculum vulgare* Mill.) were increased significantly due to an application of 120 kg ha^{-1} nitrogen in combination with 60 kg ha^{-1} phosphorus, under Junagadh (Gujrat) conditions.

Sunanda *et al.* (2014) reported that, the maximum (620.17) number of pods plant^{-1}, pod length (2.14) and number of seeds pod^{-1} (6.06) were gained by kasuri methi crop wherein, fertilizer dose was applied @ 60:25:50 kg ha^{-1} NPK + FYM (7.5 t ha^{-1}) + *Rhizobium* (1.5 t ha^{-1}) + *Azospirillum* (5 kg ha^{-1}) + PSB (5 kg ha^{-1}), under hill zone of Karnataka.

Kumar *et al.* (2015) reported that, significantly the maximum (40.7) pods plant^{-1}, pod length (10.6 cm), pod weight (0.382 g) and seeds pod^{-1} (17.2) were obtained by fenugreek plant fertilized @ 50:50:25 kg ha^{-1}, under Ajmer (Rajasthan) conditions.

Bhutia and Sharangi (2016) studied the effect of dates of sowing and soil moisture level in different growth stages and yield dynamics of fenugreek during 2012-13 and 2013-14 and found that, the maximum pods plant^{-1} (37.8 and 36.3, respectively) were produced, when the crop grown on the 2nd November. While, maximum seeds plant^{-1} (16.6 and 17.09, respectively) were obtained, when the crop grown on the 23rd November, under Mohanpur (West Bengal) conditions.

Singh *et al.* (2017^{b}) noticed that, the number of pods plant^{-1} were increased significantly due to an application of 60 kg ha^{-1} phosphorus in fenugreek, under Meerut (U. P.) conditions.

Meena *et al.* (2018) reported that, the maximum number of pods plant^{-1} (34.56), pod length (9.92 cm) and number of seeds pod^{-1} (16.12) were produced by fenugreek plants fertilized @ 50:50:25 kg ha^{-1} NPK, under Ajmer (Rajasthan) conditions.

Mitoo *et al.* (2018) conducted an experiment in fenugreek and found that, application of phosphorus 40 kg ha^{-1} and Sulphur 10 kg ha^{-1} gave the maximum (15.33) number of seeds pod^{-1}, number of pods (20.93) and test weight (9.85 g), under Bangladesh conditions.

2.2.5 Interaction effect of dates of sowing and fertilizer doses on flowering

Bhardwaj (1996) noticed that, the minimum (32.00) days to 50 per cent flowering were required by french bean crop wherein fertilizer dose applied @ 36:75:22.5 kg ha^{-1} NPK and crop sown on the date of 05th July, under Nauni (Solan) conditions.

Sheoran *et al.* (2000) observed that, minimum days were required for flower initiation in fenugreek genotypes, HM-65 and T-8, when the crop was sown on 16th December and fertilized with 90 kg phosphorus ha^{-1}, under Hisar conditions.

Nandre *et al.* (2011) found that, the interaction effect of different fertilizer dose and dates of sowing, in terms of days required for flowering were found to be non-significant, during both the years of experimentations 2004-05 and 2005-06, respectively, under Akola (M. S.) conditions.

2.2.6 Interaction effect of dates of sowing and fertilizer doses on pods per floret, seeds per pod and florets per plant

Bhardwaj (1996) revealed that, the maximum (103.90) days to seed harvest, pods plant^{-1} (18.32) were recorded by french bean crop, when fertilized @ 54:125:37.5 kg ha^{-1} NPK and crop sown on 5th June. While, the maximum (14.27 cm) pod length and seeds pod^{-1} (4.97) were obtained, when the fertilizer dose applied @ 45:100:30 kg ha^{-1} NPK and crop was sown on 20th June, under Nauni (Solan) conditions.

Nandre *et al.* (2011) noticed that, the pods plant^{-1} and seeds pod^{-1} in fenugreek (*Trigonella foenum-graecum* L.) were expressed non-significant effects due to sowing date and application of fertilizer dose, under Akola (M. S.) conditions.

Meena *et al.*, (2018) stated that, NPK applied to the fenugreek crop sown on different dates in *rabi* season of 2016, found non-significant effects on pods plant^{-1} and seeds pod^{-1}, whereas, significant effects on pod length, under Ajmer (Rajasthan) conditions.

2.3 RELATION OF SEED YIELD WITH THE DATES OF SOWING AND FERTILIZER DOSES IN KASURI METHI

2.3.1 Effect of dates of sowing on seed yield in kasuri methi

Bhardwaj (1996) reported that, the maximum (12.580 g) seed yield plant^{-1} and seed yield ha^{-1} (11.40 q ha^{-1}) was noticed when crop was sown on the date of 20th June 1996, in french bean (*Phaseolus vulgaris* L.) cultivar, Contender and Pant Anupama, under Nauni (Solan) conditions.

Sheoran *et al.* (2000) opined that, the maximum (20.80 q ha^{-1}) seed yield in fenugreek genotypes HM-65 and T-8 was gained, when the crop was sown on 1st December.

Meena *et al.* (2006) noted that, the seed yield in coriander was (*Coriandrum sutivum* L.) increased significantly with an early sowing i.e. 15th October as compared to the delayed sowing, under Udaipur (Rajasthan) conditions.

Sharma (2006) found that, significantly the highest (9.08 q ha^{-1}) seed yield ha^{-1} was obtained, with the sowing date 10th October 2005 in kasuri methi (*Trigonella corniculata* L.), under Jabalpur (M. P.) conditions.

Nandre *et al.* (2011) noted that, the per plot and per hectare seed yield of fenugreek were significantly the higher from first date of sowing i.e. 1st November, whereas successive delayed by each 15 days in sowing resulted into the significant reduction of seed yield, under Akola (M. S.) conditions.

Under Peshawar (Pakistan) conditions, in an experiment conducted on coriander, Ali *et al.* (2015) reported that, the sowing undertaken on 5th November had produced the maximum (1.485 q ha^{-1}) seed yield ha^{-1} as compared to other dates of sowing.

Anitha *et al.* (2016) studied the effect of sowing date on yield and quality of fenugreek and expressed that, significantly the maximum (14.36) pods $plant^{-1}$ and seed yield $plant^{-1}$ (09.99 g) were produced, when the crop grown on 15th October, under Godavari (A. P.) conditions.

Bhutia and Sharangi (2016) found that, the maximum seed yield $plot^{-1}$ in the year of 2012-13 and 2013-14 (515 g m^{-2} and 525 g m^{-2}, respectively) was obtained, when the crop was sown on the 2nd November, under Mohanpur (West Bengal) conditions.

Singh *et al.* (2017[a]) noticed that, significantly the highest seed yield (18.55 q ha^{-1}) was recorded by fenugreek crop sown on 10th October, under Gwalior (M.P) conditions.

Meena *et al.* (2018) recorded that, significantly the maximum (18.00 q ha^{-1}) seed yield ha^{-1}, when crop sown on 30th October, in fenugreek, under Ajmer (Rajasthan) conditions.

2.3.2 Effect of fertilizer doses on seed yield in kasuri methi

Bhardwaj (1996) reported that, the maximum seed yield $plant^{-1}$ (10.56 g) and seed yield ha^{-1} (6.38 q ha^{-1}) was produced by french bean (*Phaseolus vulgaris* L.) crop when fertilizer dose applied @ 54:125:37.5 kg ha^{-1} NPK, under Nauni (Solan) conditions.

Sheoran *et al.* (2000) noticed that, the maximum (20.80 q ha^{-1}) seed yield in fenugreek (*Trigonella foenum-graecum* L.) genotypes HM-65 and T-8 was obtained, when the phosphorus was applied @ 90 kg ha^{-1}.

Datta *et al.* (2005) found that, significantly the maximum (1.07 ha^{-1}) seed yield ha^{-1} was gained by fenugreek crop, with the application of nitrogen @ 45 kg ha^{-1}, under Mohanpur (West Bengal) conditions.

Gurjer (2005) reported that, significantly the highest (16.97 q ha^{-1}) seed yield ha^{-1} was produced by kasuri methi crop with the application of 40:40:20 kg ha^{-1} NPK + FYM 5 t ha^{-1}, under Jabalpur (M. P.) conditions.

Meena *et al.* (2006) noted that, the seed yield of coriander (*Coriandrum sutivum* L.) increased significantly with the increasing nitrogen levels i.e. 75 kg ha^{-1} as compare to low nitrogen levels, under Udaipur (Rajasthan) conditions.

Sharma (2006) found that, significantly the highest (9.08 q ha^{-1}) seed yield ha^{-1} was obtained in kasuri methi, with the fertilizer dose 40:40:20 kg ha^{-1} (100 %) NPK + 5 t ha^{-1} FYM, under Jabalpur (M. P.) conditions

Kumar and Singh (2007) opined that, an application of 50 kg ha^{-1} phosphorus resulted into the production of maximum (2.27 t ha^{-1}) seed yield in fenugreek, under IARI, New Delhi conditions.

Nandre *et al.* (2011) revealed that, an application of dose of fertilizer i.e. 37.50:18.75 kg ha^{-1} NP and 10 t ha^{-1} FYM to fenugreek crop gave significantly the highest (0.690 kg) seed yield $plot^{-1}$ and ha^{-1} (17.22 q ha^{-1}), under Akola (M. S.) conditions.

Koyani *et al.* (2012) noticed that, an application of 120 kg ha^{-1} nitrogen in combination with 60 kg ha^{-1} phosphorus had significant effect on seed yield of fennel (*Foeniculum vulgare* Mill.), under Junagadh (Gujrat) conditions.

Sunanda *et al.* (2014) reported that, the highest (4.65 q ha^{-1}) seed yield was attained when, the fertilizer dose @ 60:25:50 kg ha^{-1} NPK + FYM (7.5 t ha^{-1}) + *Rhizobium* (1.5 t ha^{-1}) + *Azospirillum* (5 kg ha^{-1}) + PSB (5 kg ha^{-1}) was applied to kasuri methi, under hill zone of Karnataka.

Ali *et al.* (2015) noted that, there was a significant increase in seed yield in coriander with application of increased dose of nitrogen. Application of nitrogen @ 60 kg ha^{-1} gave the maximum (1.62 q ha^{-1}) seed yield, under Peshawar (Pakistan) conditions.

Kumar *et al.* (2015) opined that, significantly the highest (22.49 q ha^{-1}) seed yield of fenugreek was produced, when fertilizer dose was applied @ 50:50:25 kg ha^{-1}, under Ajmer (Rajasthan) conditions.

On the basis of two years experimentations, Datta *et al.* (2017) recorded that, an application of 60 kg nitrogen, 80 kg phosphorus and 40 kg potash ha^{-1} had recorded the maximum seed yield $plot^{-1}$ (688.14 g) and yield ha^{-1} (17.20) in fenugreek, under Mohanpur (West Bengal) conditions.

Singh *et al.* (2017[b]) noticed that, the total seed yield was increased significantly due to an application of 60 kg ha^{-1} phosphorus and get 23.43 q ha^{-1} seed yield of fenugreek, under Meerut (U. P.) conditions.

Meena *et al.*, (2018) reported that, significantly the maximum (16.75 q ha^{-1}) seed yield ha^{-1} was produced, when fertilized with 50:50:25 kg ha^{-1} NPK in fenugreek, under Ajmer (Rajasthan) conditions.

Mitoo *et al.* (2018) concluded that, application of phosphorus 40 kg ha^{-1} and Sulphur 10 kg ha^{-1} increased 50.69 per cent seed yield of fenugreek over the control treatment.

2.3.3 Interaction effect of dates of sowing and fertilizer doses on seed yield in kasuri methi

Bhardwaj (1996) found that, the maximum (11.72 q ha^{-1}) seed yield ha^{-1} was obtained when fertilizer dose applied @ 54:125:37.5 kg ha^{-1} NPK and the french bean crop sown on 20th June, under Nauni (Solan) conditions.

Sheoran *et al.* (2000) conducted an experiment on influence of sowing time and phosphorus on phenology, thermal recruitment and yield of fenugreek (*Trigonella foenum-graecum* L.) genotype and expressed that, the maximum (20.80 q ha^{-1}) seed yield production in fenugreek genotypes HM-65 and T-8 was observed, when the crop was sown on 1st December with phosphorus applied @ 90 kg ha^{-1}.

Gurjer (2005) reported that, significantly the highest (16.97 t ha^{-1}) seed yield ha^{-1} was obtained with the application of 40:40:20 kg ha^{-1} NPK + FYM 5 t ha^{-1} in kasuri methi (*Trigonella corniculata* L.), grown under Jabalpur (M. P.) conditions.

Sharma (2006) found that, significantly the highest seed yield ha^{-1} (9.08 q ha^{-1}) was produced with the application of 40:40:20 kg ha^{-1} (100 %) NPK + 5 t ha^{-1} FYM. While, minimum (4.12 q ha^{-1}) seed yield ha^{-1} was gained with control treatment in kasuri methi, grown under Jabalpur (M. P.) conditions.

In the seed production experiment conducted at Akola (M. S.), Nandre *et al.* (2011) reported that, significantly the maximum $plot^{-1}$ and ha^{-1} seed yield from fenugreek variety Pusa Early Bunching which were harvested from 1st November sown crop and fertilized with 37.50:18.75 kg ha^{-1} NP along with 10 t ha^{-1} FYM.

Ali *et al.* (2015) pointed that, the seed yield (1.70 q ha^{-1}) was the maximum in coriander (*Coriandrum sutivum* L.) crop, when it was sown on 5th November and fertilized with 60 kg ha^{-1}, under Peshawar (Pakistan) conditions.

Meena *et al.*, (2018) revealed that, the maximum yield ha^{-1} (25.11 q ha^{-1}) of coriander was obtained when fertilized @ 50:50:25 kg ha^{-1} NPK to the crop sown on 30th October 2016, under Ajmer (Rajasthan) conditions.

2.4 RELATION OF SEED AND LEAVES QUALITY WITH THE DATES OF SOWING AND FERTILIZER DOSES IN KASURI METHI

2.4.1 Effect of dates of sowing on seed and leaves quality

Bhardwaj (1996) concluded that, the maximum (31.33 g) seed index, germination per cent (67.56 %) were obtained by french bean crop sown on the date of 20th June, under Nauni (Solan) conditions.

Meena *et al.* (2006) noted that, the test weight of coriander (*Coriandrun sutivum* L.) increased significantly with an early sowing i.e. 15th October as compared to the delayed sowing, under Udaipur (Rajasthan) conditions.

Nandre (2006) recorded that, significantly the maximum (1.367 mg g^{-1}) chlorophyll content, test weight (16.51 g), germination per cent (89.65 %) and seed protein content (21.10 %) were observed in the earlier sowing date i.e. 1st November during both the years of experimentations. And further, they have reported that, the chlorophyll content, test weight of seed, germination per cent and seed protein content were reduced due to delay in sowing of the crop.

Anupama (2012) reported that, significantly the maximum chlorophyll A (1.55 mg), B (0.96 mg) and total chlorophyll content (2.51 mg 100 mg^{-1}) at 60 DAS was noticed in kasuri methi, when crop was sown on 28th November 2011, with the spacing of 30 cm x 20 cm, under Arabhavi (Karnataka) conditions.

In the same year Koyani *et al.* (2012) noticed that, an application of 120 kg ha^{-1} nitrogen in combination with 60 kg ha^{-1} phosphorus had significant effect on test weight of fennel (*Foeniculum vulgare* Mill.), under Junagadh (Gujrat) conditions.

Abou-Shleel (2014) reported that, there were significant differences of sowing dates on chemical composition and active ingredients of fenugreek seeds. The sowing date 1st October was recorded the highest values of trigonelline per cent, protein per cent and seed index (0.73 %, 25.25 % and 1.45 g, respectively) of seed. Subsequently, delay in sowing resulted into reduction in trigonelline, protein content and seed index, under Egypt conditions.

Pasricha and Gupta (2014) quoted that, kasuri methi leaves and petioles consist of phenolic content in the range of (17.30µg GAE/mg) and (39.83 µg GAE/mg) while analyzing it using methanolic and aqueous extract, respectively, under Dwarka (Gujrat) conditions.

Anitha *et al.* (2016) studied the effect of sowing date on quality of fenugreek and found that, significantly the maximum chlorophyll A and B (3.43 mg g^{-1} and 1.18 mg g^{-1}, respectively), total chlorophyll (4.61 g^{-1}) and protein contenting seed (10.91 %), when the crop was sown on 15th October 2014, under Godavari (A. P.) conditions.

Bhutia and Sharangi (2016) found that, the maximum test weight in the years of 2012-13 and 2013-14 (17.6 g and 17.9 g, respectively) in fenugreek was produced when the crop sown on the 2nd November, under Mohanpur (West Bengal) conditions.

Singh *et al.* (2017[a]) found that, significantly the maximum germination per cent (97.33 %), protein content in seed (18.63 %) and highest test weight (20.20 g) were recorded in fenugreek, when the crop sown on 10th October 2015, under Gwalior (M.P) conditions.

In the same year Singh *et al.* (2017[b]) noticed that, the test weight of seed was increased significantly due to an application of 60 kg ha^{-1} phosphorus upto (7.56 g), under Meerut (U. P.) conditions.

Meena *et al.* (2018) recorded that, significantly the maximum test weight (14.95 g) in fenugreek was obtained, when crop sown on 30th October 2016, under Ajmer (Rajasthan) conditions.

2.4.2 Effect of fertilizer doses on seed and leaves quality

Bhardwaj (1996) reported that, the maximum (28.17 g) seed index, germination per cent (62.50 %) in french bean were reported, when the fertilizer dose applied at the rate of 54:125:37.5 kg ha^{-1} NPK, under Nauni (Solan) conditions.

Meena *et al.* (2006) noted that, the test weight of coriander (*Coriandrum sutivum* L.) was increased significantly with the increasing nitrogen levels i.e. 75 kg ha^{-1}, as compare to low nitrogen levels, under Udaipur (Rajasthan) conditions.

Nandre (2006) recorded that, significantly the maximum (1.68 mg g^{-1}) chlorophyll content, test weight (17.39 g), germination per cent (91.35 %) and seed protein content (21.16 %) were observed in higher level of fertilizers i.e. 37.50:18.75 kg ha^{-1} NP and 10 t ha^{-1} FYM, during both the years of experimentations. And further, they have reported that, the chlorophyll content, test weight of seed, germination per cent and seed protein content were reduced, with reduction in level of fertilizer.

Kumar and Singh (2007) opined that, an application of 50 kg ha^{-1} phosphorus had gave the maximum (14.5 g) test weight in fenugreek, under IARI, New Delhi conditions.

Anupama (2012) reported that, significantly the maximum chlorophyll A (1.61 mg), B (1.38 mg) and total chlorophyll content (2.99 mg 100 mg^{-1}) at 60 DAS was obtained, when nitrogen was given @ 80 kg ha^{-1}, in kasuri methi (*Trigonella corniculata* L.), under Arabhavi (Karnataka) conditions.

Singh *et al.* (2012) studied the effect of inorganic and bio-fertilizers and found that, the significant increase in per cent protein content in leaves of kasuri methi. They reported the maximum protein content in leaves with 100 % RDF and Rhizobium + PSB, under Jobner (Rajasthan) conditions.

Sunanda *et al.* (2014) reported that, the maximum (21.44 mg 100 mg^{-1}) chlorophyll content at 60 DAS, test weight (1.75 g) and protein content of the seed (21.33 %) were reported, when, the fertilizer dose @ 60:25:50 kg ha^{-1} NPK + FYM (7.5 t ha^{-1}) + *Rhizobium* (1.5 t ha^{-1}) + *Azospirillum* (5 kg ha^{-1}) + PSB (5 kg ha^{-1}) in kasuri methi, under hill zone of Karnataka.

Anupama *et al.* (2017) reported that, significantly increasing trend of the chlorophyll content in leaves with increasing the rate of nitrogen dose. They reported that, the maximum chlorophyll A, B and total chlorophyll content (1.61, 1.38 and 2.99 mg 100 mg^{-1}, respectively) were obtained, when nitrogen applied @ 80 kg ha^{-1}, under Arabhavi (Karnataka) conditions.

In the same year Datta *et al.* (2017) observed that, an application of 60 kg nitrogen, 80 kg phosphorus and 40 kg potash ha^{-1} had produced the maximum (15.24 g) test weight in fenugreek, under Mohanpur (West Bengal) conditions.

Meena *et al.* (2018) reported that, significantly the maximum (14.11 g) test weight in fenugreek was obtained, when fertilizer dose applied @ 50:50:25 kg ha^{-1} NPK, under Ajmer (Rajasthan) conditions.

2.4.3 Interaction effect of dates of sowing and fertilizer doses on seed and leaves quality in kasuri methi

Bhardwaj (1996) revealed found that, the maximum (31.71 q ha^{-1}) seed index and germination per cent (70.00 %) were observed in french bean, when fertilizer dose applied @ 54:125:37.5 kg ha^{-1} NPK and the crop was sown on 20th June, under Nauni (Solan) conditions.

Nandre (2006) recorded that, the maximum chlorophyll content, test weight, germination per cent and seed protein content were observed in the earlier sowing date with highest fertilizer level i.e. 1st November and 37.50:18.75 kg ha^{-1} NP and 10 t ha^{-1} FYM, during both the years of experimentations.

Anupama (2012) found that, significantly the highest chlorophyll A, B and total chlorophyll content at 60 DAS (1.80 mg, 1.67 mg, 3.48 mg 100 mg^{-1}, respectively) were noticed with the application of nitrogen dose @ 80 kg ha^{-1} and spacing of 30 cm x 20 cm was obtained in kasuri methi (*Trigonella corniculata* L.), sown on the date 28th November, under Arabhavi (Karnataka) conditions.

Jawarkar (2017) conducted the experiment on leaves production of kasuri methi and reported that, the non-significant effect of spacing and pinching on oleoresin and trigonelline content (%), under Akola (M. S.) conditions.

2.5 RELATION OF SOIL AND PLANT NUTRIENT STATUS WITH THE DATES OF SOWING AND FERTILIZER DOSES IN KASURI METHI

Gurjer (2005) studied that, an application of 40:40:20 kg ha^{-1} NPK + FYM 5 t ha^{-1} in kasuri methi, (*Trigonella corniculata* L.) grown, under Gwalior (M. P.) conditions and found Initial status of soil i.e. nitrogen, phosphorus and potassium were (196.52, 15.05, 233.70 kg ha^{-1}, respectively). Significantly the maximum total nitrogen, phosphorus and potassium (99.60, 18.69, 49.19 kg ha^{-1}, respectively) uptake by the plant however, available nitrogen, phosphorus and potassium of soil (213.25, 28.70, 240.80 kg ha^{-1}, respectively)

Nandre (2006) recorded that, the maximum nutrient content of plant and total uptake of nutrients by the plant were observed in the earlier sowing date with highest fertilizer level i.e. 1st November and 37.50:18.75 kg ha^{-1} NP and 10 t ha^{-1} FYM during both the years of experimentations. However, non-significant interaction effect of sowing date and nutrient management was observed.

Sharma (2006) found that, significantly the highest uptake of nitrogen, phosphorus and potash (60.07, 11.78, 30.53 kg ha^{-1}, respectively) were noticed, with the application of 40:40:20 kg ha^{-1} (100 %) NPK + 5 t ha^{-1} FYM in kasuri methi, grown under Jabalpur (M. P.) conditions.

Anupama (2012) found that, kasuri methi crop sown on 28th November with the spacing of 30 cm x 20 cm exhibited significantly the maximum total dry matter at 45, 60 and 75 DAS (4.47, 8.32 and 14.03 g plant^{-1}, respectively) and the maximum (1.05 %) nitrogen content in plant, nitrogen uptake by plant and available nitrogen in soil (34.88, 213.20 kg ha^{-1}, respectively), under Arabhavi (Karnataka) conditions.

Sunanda *et al.* (2014) reported that, significantly the maximum total dry matter at 30, 60, 90 DAS (0.76, 7.14, 15.45 g plant^{-1}, respectively) was obtained, when fertilizer dose given @ 60:25:50 kg ha^{-1} NPK + FYM (7.5 t ha^{-1}) + *Rhizobium* (1.5t ha^{-1}) + *Azospirillum* (5 kg ha^{-1}) + PSB (5 kg ha^{-1}) in kasuri methi, under hill zone of Karnataka.

Babaleshwar and Shetty (2017) observed that, significantly the highest total nitrogen, phosphorous and potash uptake by the treatment T_7 (60:25:50 kg ha^{-1} NPK + FYM 7.5 t ha^{-1}) + *Rhizobium* (1.5 kg ha^{-1}) + *Azospirillum* + PSB (5 kg ha^{-1}) in kasuri methi were produced, under hill zone of Bagalkot, Karnataka.

2.6 ECONOMIC RELATION OF DATES OF SOWING AND FERTILIZER DOSES ON SEED PRODUCTION IN KASURI METHI

Bhardwaj (1996) noticed that, the maximum gross monetary return (58400 Rs. ha^{-1}), net monetary return (33764.39 Rs. ha^{-1}) and highest B:C ratio (1.37) were recorded in french bean, when the crop was fertilizer @ 54:125:37.5 kg ha^{-1} NPK and crop sown on 20th June, under Nauni (Solan) conditions.

Nandre and Sahane (2011) reported that, the maximum gross monetary return (52211 Rs. ha^{-1}), net monetary return (35984 Rs. ha^{-1}) were obtained, when the crop grown on 1st November (S_1). Whereas, maximum gross monetary return (62271 Rs. ha^{-1}), net monetary return (49229 Rs. ha^{-1}) were noticed, when fertilizer dose applied @ 50:25:0 kg ha^{-1} NPK (F_2) and got highest B:C ratio (5.59) in (S_1F_2) for the seed production in fenugreek (*Trigonella foenum-graecum* L.), under Akola (M. S.) conditions.

Anupama (2012) reported that, significantly the highest gross monetary return, net monetary return and cost benefit ratio (371400 Rs. ha^{-1}, 261969 Rs. ha^{-1} and 3:3.39, respectively) for herb yield was obtained with an application of nitrogen @ 80 kg ha^{-1} and spacing of 20 cm x 10 cm in kasuri methi (*Trigonella corniculata* L.) sown on 28th November, under Arabhavi (Karnataka) conditions.

Koyani *et al.* (2012) found that, an application of 120 kg ha^{-1} nitrogen and 60 kg ha^{-1} phosphorus had significant effect on the gross monetary return, net monetary return and the highest B:C ratio in fennel (*Foeniculum vulgare* Mill.), under Junagadh (Gujrat) conditions.

Singh *et al.* (2012) studied the effect of inorganic and bio-fertilizers and found that, the significant increase in net return and B:C ratio in kasuri methi. They quoted that, the maximum (154322 Rs. ha^{-1}) net return and B:C ratio (2.98) was obtained with 100 % RDF and Rhizobium + PSB, under Jobner (Rajasthan) conditions.

Godara *et al.* (2013) reported that, the maximum (140140 Rs. ha^{-1}) gross monetary return, net monetary return (105522 Rs. ha^{-1}) and B:C ratio (4.05), when phosphorus and Sulphur were applied @ 60 kg ha^{-1} and 45 kg ha^{-1}, respectively, in kasuri methi for the leaves purpose, under Ajmer (Rajasthan) condition.

Kumar *et al.* (2015) reported that, significantly the maximum (73660.6 Rs. ha^{-1}) gross monetary return, net monetary return (51091.56 Rs. ha^{-1}) and B:C ratio (2.3) were obtained, when fertilizer dose applied @ 50:50:25 kg ha^{-1} NPK in fenugreeek, under Ajmer (Rajasthan) conditions.

Meena *et al.* (2018) revealed that, significantly the maximum (109632.71 Rs. ha^{-1}) net monetary return and B:C ratio (3.22) were obtained, when the fertilizer dose applied @ 50:50:25 kg ha^{-1} NPK to the fenugreek crop sown on 30th October, under Ajmer (Rajasthan) conditions.

3

CHAPTER

MATERIAL AND METHODS

The present investigation entitled "Effect of sowing time and fertilizer levels on seed yield of kasuri methi (*Trigonella corniculata* L.)" was undertaken at Chilli and Vegetable Research Unit (CVRU), Dr. Panjabrao Deshmukh Krishi Vidyapeeth, Akola, during *rabi* seasons of 2016-17 and 2017-18. The details of material used and methods adopted during the course of investigation are given in this chapter, under the appropriate headings and sub-headings.

3.1 EXPERIMENTAL SITE

The present investigation was carried out during *rabi* seasons of 2016-17 and 2017-18, at Chilli Vegetable Research Unit Research Unit, Dr. Panjabrao Deshmukh Krishi Vidyapeeth, Akola (M. S.).

3.2 CLIMATE AND WEATHER CONDITIONS

Akola comes under sub-tropical zone and is situated at a latitude of 20°70′ N and longitude of 77°03′ E. The altitude of the place is 282 m from mean sea level. The mean annual precipitation on the basis of last fifteen years is 800 mm, which receives almost from South-West monsoon, during June to October. The mean annual minimum and maximum temperatures are 13.80°C and 34.1°C, respectively. The humidity ranges from 29.75 per cent in summer and 61.23 per cent in rainy season. Thus Akola has hot and dry summer with moderate cold winter. The details of meteorological data were recorded monthly in respect of rainfall, maximum and minimum temperatures and relative humidity at Agro-meteorological observatory, Dr. Panjabrao Deshmukh Krishi Vidyapeeth, Akola for both the years of study 2016-17 and 2017-18 have been furnished in Appendix-I and Appendix-II, respectively.

3.3 SOIL

The land used under the experimental layout was fairly uniform with gentle slope. The soil was medium black with uniform texture. The plot had a good drainage. On the experimental field, vegetables nursery was grown in the previous season of the experiments during both the years of experimentations. To know the chemical properties of soil of experimental plot, the representative soil samples were collected from 15 to 20 spots after scrapping the surface litter. A uniform thin slice of the soil from surface to a plough depth of 15 to 20 cm was collected before the start of an experiment (Seetharaman *et al.* 1994). Details of chemical and physical analysis of the soil of experimental plot are furnished in the tabular form.

Physico-chemical properties of soil

Sr. No	Ingredients	Quantity	Estimation methods
1	Sand	27.30 %	International pipette method (Piper, 1966)
2	Silt	26.21 %	
3	Clay	49.71 %	
4	Soil pH	7.64	Glass electrode pH meter (Jackson, 1967)
5	Available nitrogen	190.34 kg ha^{-1}	Alkaline permanganate method (Subbiah and Asija, 1956)
6	Available phosphorus	10.24 kg ha^{-1}	Olsen's method (Olsen et al., 1954)
7	Available potassium	368.72 kg ha^{-1}	Neutral normal ammonium acetate method (Jackson, 1973)

The fertility status of the soil of experimental plot is presented in above table which revealed that, the soil was clayey in texture, slightly alkaline in nature, low in available nitrogen, very low in phosphorus and very high in available potassium.

3.4 MATERIALS REQUIRED

3.4.1 Planting material

The seeds of kasuri methi variety 'Rajendra Abha' was collected from Dr. Rajendra Prasad Central Agricultural University Pusa, Samastipur, Bihar, India, during the first experimental trial in the year 2016-17 and same harvested seeds was used for second trial conducted in the year 2017-18.

3.4.2 Preparatory tillage

An experimental land was ploughed and cross harrowing's were given in both the years to bring the soil to the fine tilth. The clods were crushed by the clod crusher to make the soil loose and friable. The weed growth and stubbles of previous crop were picked up and cast it. The field was laid out with flat bed of desired dimensions before 3-4 days of sowing of seeds during both the years of the experimentations.

3.4.3 Seed sowing

Before the seed sowing, seeds were pre-treated with thirum @ 2 g kg^{-1} of seed. At the every sowing time, only plots to sow were irrigated 2-3 days before the sowing date, during both the years of experiments, so as to facilitate easy sowing. Before sowing, seeds were mixed with sand to maintain proper seed rate and spacing. The sowing was undertaken on the flat bed by dibbling method, as per sowing date keeping 1-1.5 cm soil depth. The seeds were sown on the moist beds by maintaining the spacing of 30 cm x 20 cm, during both the years of experimentations 2016-17 and 2017-18.

3.4.4 Fertilizer application

The treatment-wise fertilizer requirement was worked out and then half dose of nitrogen in the form of urea and full dose of phosphorus in the form of single super phosphate, full dose of potassium in the form of muraite of potash were applied at the time of planting. Remaining half dose of nitrogen was applied at 30 days after planting as top dressing.

3.4.5 Irrigation

The experimental field was irrigated at 6 to 10 days interval throughout the growing season of the crop. The first irrigation was given 2-3 days before sowing and second irrigation was given after germination (3-4 days after sowing). All thirty two irrigations were given to the seed crop of kasuri methi, as given in the tabular form below:

Irrigation no.	Dates of irrigation	
	2016-17	2017-18
1	08.10.2016	07.10.2017
2	13.10.2016	12.10.2017
3	22.10.2016	21.10.2017
4	28.10.2016	27.10.2017
5	04.11.2016	05.11.2017
6	11.11.2016	11.11.2017
7	18.11.2016	17.11.2017
8	24.11.2016	23.11.2017
9	02.12.2016	01.12.2017
10	07.12.2016	08.12.2017
11	13.12.2016	14.12.2017
12	20.12.2016	21.12.2017
13	27.12.2016	28.12.2017
14	03.01.2017	02.01.2018

15	10.01.2017	11.01.2018
16	17.01.2017	18.01.2018
17	24.01.2017	23.01.2018
18	29.01.2017	28.02.2018
19	04.02.2017	04.02.2018
20	09.02.2017	10.02.2018
21	15.02.2017	16.02.2018
22	21.02.2017	22.02.2018
23	27.02.2017	28.02.2018
24	02.03.2017	03.03.2018
25	08.03.2017	07.03.2018
26	12.03.2017	11.03.2018
27	18.03.2017	19.03.2018
28	24.03.2017	25.03.2018
29	30.03.2017	29.03.2018
30	04.04.2017	05.04.2018
31	09.04.2017	10.04.2018
32	14.04.2017	15.04.2018

3.4.6 Weeding

Two hand weedings were given to keep the plot free from weeds. The first hand weeding was undertaken at 15 days after sowing of seeds, second hand weeding was given at 45 days after first weeding, during both the years of experimentation.

3.4.7 Thinning of plants

Thinning was done at 20 days after sowing to maintain plant populations, as well as to retain the 5 to 6 plants at every bridge, as per the treatment.

3.4.8 Plant protection

There was no any disease or pest incidence observed throughout experiment during both the years of 2016-17 and 2017-18.

3.4.9 Harvesting of seeds

The crop was harvested as per maturity of pods. The irrigation was stopped, when the pods were fully matured i.e. when the pod turned green to brown. The harvesting was done before the drying of the stem and florets but after the drying of pods. The harvesting was done early in the morning (It facilitates to prevent the drop of pods at the time of harvesting), manually with cutting of the whole stem just above the ground level with the help of sickle. The harvesting was done separately with individual treatment and dried on clean threshing floor. The threshing and cleaning of seeds was done manually from each of the treatment plot of the experiments during both the years of the experimentations.

3.5 METHODS ADOPTED

3.5.1 Experimental details

1.	Location	:	Chilli and Vegetable Research Unit Dr. PDKV Akola
2.	Crop	:	Kasuri methi (*Trigonella corniculata.* L.
3.	Variety	:	Rajendra Abha
4.	Family	:	Leguminaceae
5.	Spacing	:	Row to row 30 cm Plant to plant 20 cm
6.	Years	:	2016-2017 and 2017-18
7.	Sowing season	:	Rabi
8.	Treatments	:	Factor A- Dates of sowing Factor B- Fertilizer doses (N:P:K.
9.	Treatment combinations	:	24
10.	Replications	:	Three (3.
11.	Design	:	Factorial Randomized Block Design (FRBD.
12.	Plot size	:	(a) Gross plot size : 2.25 m x 2.25 m (b) Net plot size : 2.10 m x 2.00 m
13.	Plants per plot	:	70 (7 rows : 10 plants/ row.
14.	Total experimental area	:	460 m^2

3.5.2 Treatment details

Factor A- Dates of sowing

1. D_1- 10th October
2. D_2- 30th October
3. D_3- 20th November
4. D_4- 10th December
5. D_5- 30th December
6. D_6- 20th January

Factor B- Fertilizer doses (N:P:K)

1. F_1- 40:40:20 kg ha^{-1}
2. F_2- 40:50:25 kg ha^{-1}

3. F_3- 50:40:20 kg ha^{-1}
4. F_4- 50:50:25 kg ha^{-1}

3.5.3 Treatment combinations

Treatments	Combination	Treatments	Combination
T_1	D_1F_1	T_{13}	D_4F_1
T_2	D_1F_2	T_{14}	D_4F_2
T_3	D_1F_3	T_{15}	D_4F_3
T_4	D_1F_4	T_{16}	D_4F_4
T_5	D_2F_1	T_{17}	D_5F_1
T_6	D_2F_2	T_{18}	D_5F_2
T_7	D_2F_3	T_{19}	D_5F_3
T_8	D_2F_4	T_{20}	D_5F_4
T_9	D_3F_1	T_{21}	D_6F_1
T_{10}	D_3F_2	T_{22}	D_6F_2
T_{11}	D_3F_3	T_{23}	D_6F_3
T_{12}	D_3F_4	T_{24}	D_6F_4

3.5.4 Layout of experiment

An experiment was conducted in Factorial Randomized Block Design with twenty four treatment combinations, which were replicated thrice. The allotment of treatments to the various plots were done randomly in each replication. The site of the experiment was same for both the years of an experimentation. The experiment was conducted at same site with same randomization for both the years and it is depicted in Fig. 1.

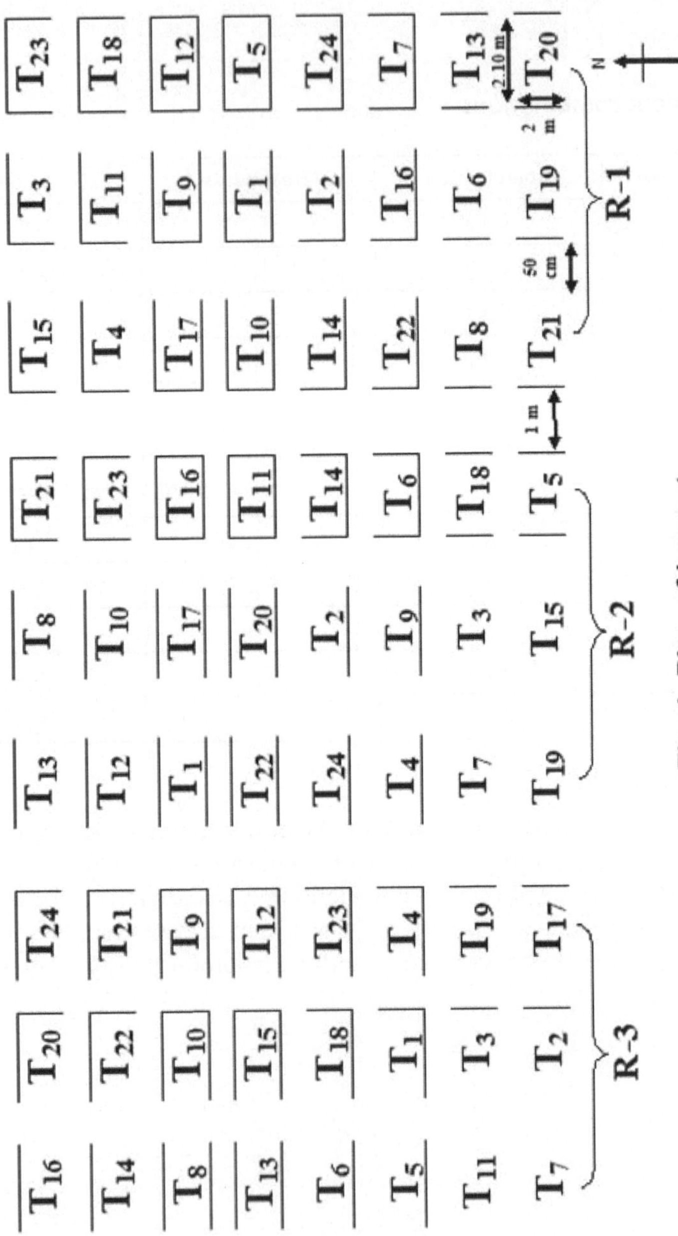

Fig. 1. Plan of Layout

3.6 OBSERVATIONS

For recording biometric observations on growth, yield and seed quality, five plants were selected at randomly from each plot. The selected plants were labelled and the observations on height, number of branches and number of leaves per plant were recorded periodically. First observation was taken 30 days after seed sowing. Subsequent observations were recorded at an interval of 30 days till harvesting stage of the crop. The following observations were recorded.

Table 1: Details of observation recorded in the field and laboratory

Sr. No	Particular	Frequency	Days after sowing
I.	Growth observations		
1.	Plant height (cm)	4	30 days interval and at last harvest
2.	No. of branches plant^{-1}	4	30 days interval and at last harvest
3.	No. of leaves plant^{-1}	4	30 days interval and at last harvest
II.	Yield attributes and yield		
1.	Days to flower initiation	1	At last harvest
2	Days to 50 % flowering	1	
3.	Number of florets plant^{-1}	1	
4.	Number of pods floret^{-1}	1	
5.	Number of seeds pod^{-1}	1	
6.	Test weight (g)	1	
7.	Seed yield plant^{-1} (g)	1	
8.	Seed yield plot^{-1} (g)	1	
9.	Seed yield ha^{-1} (q)	1	
10.	Total dry matter (%)	1	
III.	a) NPK status of soil	1	Before planting
			After harvest
	b) Uptake of NPK by plant	1	After harvest
IV.	Quality parameters		
1.	Chlorophyll (mg/100mg)	1	60 DAS
2.	Oleoresins (%)	1	After harvest
3.	Trigonelline (%)`	1	
4.	Protein content (%)	1	
5.	Germination percentage (%)	1	
V.	Incidence of Powdery mildew (%)	4	30 days interval and at last harvest
VI.	Economic studies		
1.	Gross monetary return (Rs. / ha)	1	After harvest
2.	Net monetary return (Rs. / ha)	1	
3.	B : C ratio	1	

3.6.1 Growth parameters

1. Plant height (cm)

The height of plant was measured from ground level to the growing tip of five observational plants and after computing the mean, it was recorded as height of plant in centimeter (cm). The height was recorded at 30, 60, 90 and at last harvesting stage.

2. Number of branches per plant

The number of branches per plant were counted at 30, 60, 90 and at last harvesting stage by taking an actual count and after calculating an average of five observational plants, it was reported as branches per plant.

3. Number leaves per plant

Numbers of trifoliate leaves of five observational plants were counted and after computing the mean, it was recorded as leaves per plant. The leaves per plant were recorded at 30, 60, 90 and at last harvesting stage.

3.6.2 Yield attributes and yield

1. Days to flower initiation

The date of first flower initiation in each treatment plot was noted and the days required for flower initiation were calculated from the date of planting and recorded accordingly.

2. Days to 50 per cent flowering

The date of 50 per cent flowering in each treatment plot was recorded and average days required for 50 per cent flowering were calculated from the date of sowing and recorded accordingly.

3. Number of florets per plant

The actual count of total number of florets per plant were taken just before the harvesting stage and after calculating an average of five observational plants, it was reported as number of florets per plant.

4. Number of pods per floret

The actual count of total number of pods per floret were taken before the harvesting stage (when pod setting is completed) and after calculating an average of five observational plants, it was reported as number of pods per floret.

5. Number of seeds per pod

The total number of seeds per pod were counted after harvesting of crop and after calculating an average of five observational plants, it was reported as number of seeds per pod.

6. Test weight (g)

Randomly, 1000 kasuri methi seeds from each treatment were taken out. Then weight of such kasuri methi seeds was undertaken with physical weighting balance. Then, this weight was recorded in grams as test weight.

7. Seed yield per plant (g)

The seeds of the each observational plant were harvested separately, after threshing the weight of the seeds of individual plants were taken with the help of physical weighting balance. Mean of the five observational plants were presented as seed yield per plant in grams.

8. Seed yield per plot (g)

The seeds of the individual plots were harvested and threshed out separately. Then weight of the threshed and cleaned seeds from each plot were taken with the help of physical weighing balance. Seed yield per plot was calculated and recorded in grams.

9. Seed yield per hectare (q)

The seed yield per hectare in quintal was calculated from the available value of seed yield per plot in Kilograms, by arithmetic calculations using hectare factor. The calculated value was recorded accordingly in quintals.

10. Total dry matter (%)

Five randomly selected plants were taken from each treatment to determine the dry weight of plants. The plants were dried in hot air oven at 70 ± 1 ^{0}C for 72 hours. The weight of such dried material was undertaken on physical weighing balance in grams. Then, with the help of fresh weight of kasuri methi plant the dry matter per cent was calculated and recorded accordingly.

3.6.3 Quality parameters

1. Chlorophyll content (mg/100 mg)

Chlorophyll content of leaf was analyzed by collecting the healthy, fully opened matured leaves from the centre portion of the plants at 60 DAS. Total chlorophyll content of leaf tissue were determined by non-destructive method of chlorophyll estimation using dimethyl sulphoxide (DMSO) as suggested by Shoaf and Lium (1976).

The chlorophyll from leaves was calculated from the equations given below and expressed in mg/100g.

$$\text{Total chlorophyll (mg/100 mg)} = 20.2\ (\text{OD at 663 nm}) - 8.02\ (\text{OD at 645 nm}) \times \frac{V}{100 \times W}$$

where,

OD = Optical density

V=Final volume i.e. 5 ml of DMSO

W= Weight of fresh leaves (100 g sample)

2. *Oleoresin content (%)*

The estimation of oleoresin content from the powder of seed of kasuri methi was carried out by using Soxlet apparatus method by heating with electric coils of 60 watts. The sample of 2 g of each treatment was used for oleoresin estimation by using methanol as a solvent at 65 ^{0}C boiling points. The difference in observations of the empty weight of flask and after estimation was noted for each of the treatment and it was recorded in percentage.

The amount of oleoresin content from seed was calculated by using following formula:

$$\text{Oleoresin (\%)} = \frac{W_2 - W_1}{W} \times 100$$

where,

W = Sample weight

W_1 = Initial beaker weight

W_2 = Final beaker weight

3. *Trigonelline content (%)*

Trigonelline content of seed was estimated from the dry sample and extracted by using methanol and supernantant *i.e* methanolic extract of kasuri methi collected. The quantity of trigonelline was determined by high performance liquid layer chromatography (HPLC) as suggested by Rongjie *et al.* (2010).

4. *Protein content (%)*

The nitrogen content of the seeds was assessed by Kjeldahl method using Pelican Kelplus equipment. Crude protein was calculated by multiplying nitrogen content with a factor 6.25 (Anon., 1990).

$$\text{Protein (\%)} = \frac{\text{(Titrate value-Blank)} \times \text{N of } H_2SO_4 \times 0.014 \times 6.25 \times 100}{\text{Sample weight}}$$

5. *Germination per cent (%)*

The laboratory germination test was conducted as per the ISTA rules (Anon., 1985) by adopting "between paper method" in four lots. In each lot, 100-seeds were drawn at random on germination paper. Then such seeds were kept in germinator maintaining 20 to 30 ^{0}C alternate temperature along with 90 ± 2 per cent relative humidity. The first and second counts were taken on 7th and 15th day from the date of keeping the seeds in germination chamber, respectively. The number of normal seedlings were counted and expressed as germination percentage with the help of following formula.

$$\text{Germination percentage} = \frac{\text{Number of seed germinated}}{\text{Total number of seed placed for germination}} \times 100$$

Means of all the lots of kasuri methi seeds, kept for germination were considered to obtain germination percentage and recorded accordingly.

3.6.4 Incidence of Powdery mildew (%)

The actual count of the total infected plants of powdery mildew from the plot was undertaken periodically. Arithmetically converted and presented as incidence of the powdery mildew in per cent infestation, on the observation sheet.

3.6.5 NPK status

1. NPK status of soil

i. Available nitrogen

Available nitrogen was determined by alkaline permanganate (0.32% $KMnO_4$) method (Subbiah and Asija, 1956) and recorded in Kg ha^{-1}.

ii. Available phosphorus

Available phosphorus was determined by extracting the soil P_2O_5 with 0.5M $NaHCO_3$ at pH 8.5. The soil: extractant ratio was 1:20 and shaking time was 30 minutes. Phosphorus in the extract was determined colorimetrically by using Spectrophotometer at 420 nm wavelength as outlined by Olsen *et al.* (1954) and recorded in kg ha^{-1}.

iii. Available potassium

It was estimated on Systronics Flame Photometer-128 using neutral-normal-ammonium acetate (NH_4OAc, pH 7.0) as per procedure given by Jackson (1973) and recorded in Kg ha^{-1}.

2. NPK content in kasuri methi crop

i. Total nitrogen (%)

The plant samples were digested with conc. H_2SO_4 and the total nitrogen content in kasuri methi plant was determined by Kjelplus apparatus (Tandon, 1993) and recorded accordingly in per cent.

ii. Total phoshopus (%)

For determination of phosphorus, 1.0 g of kasuri methi plant sample was digested with di-acid mixture (HNO_3 + $HClO_4$) with the ratio 9:4. The yellow colour was developed with combined HNO_3 and vanadomolybdate reagent. Phosphorus was determined colorimetrically by using spectrophotometer at 420 nm wavelength (Chopra and Kanwar, 1978) and recorded accordingly in per cent.

iii. Potassium (%)

For determination of potassium, 1.0 g of kasuri methi plant sample was digested with di-acid mixture (HNO_3 + $HClO_4$) with the ratio 9:4 and it was estimated flame photometrically by feeding diluted diacid digested solution duly diluted 10 times (Piper 1966)) and recorded accordingly in per cent.

3. NPK uptake by kasuri methi crop (kg ha^{-1})

The nitrogen, phosphorus and potassium uptake by kasuri methi crop was calculated by multiplying its per cent concentration with dry matter of average kasuri methi plant. The calculated figure came out through desired formulas would then expressed in Kg ha^{-1} as an uptake of nitrogen, phosphorus and potassium by the kasuri methi crop, respectively.

3.6.6 Economic studies of treatments

1. Gross monetary return (Rs. ha^{-1})

From the total yield of each treatment plot, the gross monetary return was worked out on the basis of selling prices of the produce. Then, it was recorded accordingly in Rs. ha^{-1}.

2. Net monetary return (Rs. ha^{-1})

From the gross monetary return of each treatment plot, the expenditure incurred on the same treatment was deducted. Accordingly, the treatment wise net monetary return was worked out. Then, it was recorded accordingly in Rs. ha^{-1}.

3. B:C ratio

From the value of gross monetary return of each treatment plot and expenditure incurred on the same treatment, cost benefit ratio was worked out and recorded accordingly.

3.7 STATISTICAL METHODS

The data obtained on various observations during the course of investigation was statistically analyzed as per the methods suggested by Panse and Sukhatme (1985).

3.8 PLACE/ DURATION/ SEASONS OF EXPERIMENTS

The field experiment was carried out at Chilli and Vegetable Research Unit (CVRU), Dr. Panjabrao Deshmukh Krishi Vidyapeeth, Akola, during *rabi* seasons of the years 2016-17 and 2017-18.

Plate 1. General view of experimental plot of the year 2016-17

Plate 2. General view of experimental plot of the year 2017-18

4 CHAPTER

RESULTS AND DISCUSSION

The results of the experiment entitled "Effect of sowing time and fertilizer levels on seed yield of kasuri methi (*Trigonella corniculata* L.)" was carried out at Chilli and Vegetable Research Unit (CVRU), Dr. Panjabrao Deshmukh Krishi Vidyapeeth, Akola during *rabi* seasons of the years 2016-17 and 2017-18 are presented and discussed below in this chapter.

Objectives of the study are:

1. To study the different sowing time and fertilizer levels for seed production of kasuri methi.
2. To find out best treatment combination of sowing time and fertilizer level for seed production of kasuri methi.

The observations which were recorded on the various parameters governing to growth, seed yield, seed quality and NPK status were statistically analyzed and the results obtained from experimentation are presented and discussed in this chapter under appropriate headings and sub-headings. Also an effort were made to depict the results graphically wherever possible. Similarly, the results are discussed here in the light of available literature. An attempts have also been made to provide an appropriate reasoning for the results obtained during the course of the investigation.

A. GROWTH OBSERVATIONS

4.1 PLANT HEIGHT (CM)

The data in respect of height of plant in kasuri methi at an interval of 30, 60, 90 days and at the time of last harvest, during the years 2016-17 and 2017-18 was influenced by the sowing dates and fertilizer levels are presented in Table 2 (a), (b) and depicted in Fig. 2 and 3.

4.1.1 Effect of sowing dates

The data presented in Table 2 (a) revealed that, the height of the kasuri methi plant was influenced significantly by the sowing dates and it was increased subsequently at all the growth stages including the last harvesting stage during both the years of 2016-17 and 2017-18 experimentations.

During the year 2016-17, at the stage of 30 DAS, the treatment D_1 (10th October) was recorded the maximum (7.52 cm) plant height and which was significantly superior over rest of the treatments. Whereas, the treatment D_6 (20th January) recorded minimum plant height (5.45 cm).

Similarly, the sowing date 10th October (D_1) was recorded the maximum plant height (7.82 cm) and which was significantly superior over rest of the treatments at the stage of 30 DAS, during the year 2017-18. However, the sowing date 20th January (D_6) was recorded minimum plant height (5.14 cm).

In the year 2016-17, significantly the maximum height of kasuri methi plant (26.59 cm) was seen in the sowing date 10th October (D_1), at the stage of 60 DAS and this treatment was found to be at par with the treatment D_2 (26.08 cm). However, significantly the minimum plant height (18.86 cm) was recorded in the sowing date 20th January (D_6).

At the stage 60 DAS, during the year 2017-18, the treatment D_1 was recorded the maximum plant height (27.25 cm) and which was significantly superior over rest of the treatments. Whereas, significantly the minimum plant height (20.42 cm) was recorded in the treatment D_6.

During the year 2016-17, the maximum plant height (56.79 cm) was noted in the treatment D_1, at the stage of 90 DAS and which was found to be at par with the treatment D_2 (56.56 cm). Whereas, the minimum plant height was recorded in the treatment D_6 (46.72 cm).

Similarly, at the stage of 90 DAS, during the year 2017-18, the treatment D_1 was recorded the maximum plant height (59.73 cm) and which was at par with treatment D_2 (58.96 cm). However, the minimum plant height (42.85 cm) was recorded in the treatment D_6.

At the stage of last harvest, during the year 2016-17, significantly the maximum plant height (60.05 cm) was recorded in the treatment D_1 and

Table 2 (a). Effect of dates of sowing and fertilizer doses on plant height (cm) in kasuri methi

Treatments	Plant height (cm)											
	30 DAS			60 DAS			90 DAS			At last harvest		
	2016-17	2017-18	Pooled	2016-17	2017-18	Pooled	2016-17	2017-18	Pooled	2016-17	2017-18	Pooled
Dates of sowing (D)												
D_1 - 10th October	7.52	7.82	7.67	26.59	27.25	26.92	56.79	59.73	58.26	60.05	62.52	61.29
D_2 - 30th October	7.26	7.48	7.37	26.08	26.13	26.10	56.56	58.96	57.76	59.49	61.35	60.42
D_3 - 20th November	6.87	6.92	6.90	23.08	24.05	23.57	53.98	52.43	53.20	57.89	57.62	57.75
D_4 - 10th December	6.52	6.36	6.44	21.61	24.01	22.81	52.01	47.62	49.82	56.01	53.79	55.12
D_5 - 30th December	6.06	5.64	5.85	18.93	21.49	19.71	48.64	42.94	45.79	54.84	48.38	52.14
D_6 - 20th January	5.45	5.14	5.30	18.86	20.42	19.64	46.72	42.85	44.79	53.32	46.94	49.38
'F' test	Sig.	Sig.	Sig.	Sig.	Sig.	Sig.	Sig.	Sig.	Sig.	Sig.	Sig.	Sig.
SE(m) ±	0.07	0.07	0.05	0.31	0.27	0.22	0.24	0.35	0.24	0.24	0.44	0.25
CD at 5%	0.20	0.18	0.14	0.88	0.78	0.63	0.68	0.98	0.69	0.69	1.24	0.71
Fertilizer dose (F)												
F_1 - 40:40:20 kg ha^{-1} NPK	6.44	6.32	6.39	21.84	23.14	22.49	51.48	49.78	50.63	56.24	54.05	55.14
F_2 - 40:50:25 kg ha^{-1} NPK	6.54	6.47	6.51	22.91	23.93	23.42	52.86	51.07	51.96	57.14	55.46	56.30
F_3 - 50:40:20 kg ha^{-1} NPK	6.66	6.63	6.65	22.19	23.58	22.88	52.12	50.40	51.26	56.70	54.86	55.78

F_4 - 50:50:25 kg ha^{-1} NPK	6.80	6.82	6.81	23.16	24.25	23.71	53.35	51.77	52.56	57.66	56.03	56.84
'F' test	Sig.	Sig.	Sig.	Sig.	Sig.	Sig.	Sig.	Sig.	Sig.	Sig.	Sig.	Sig.
SE(m) ±	0.06	0.05	0.04	0.25	0.22	0.18	0.20	0.28	0.20	0.20	0.36	0.20
CD at 5%	0.16	0.15	0.12	0.72	0.64	0.51	0.56	0.80	0.57	0.57	1.01	0.58
Interaction (D X F)												
'F' test	NS	NS	NS	Sig.	Sig.	Sig.	Sig.	Sig.	Sig.	Sig.	Sig.	Sig.
SE(m) ±	0.14	0.13	0.10	0.62	0.55	0.44	0.48	0.69	0.49	0.49	0.87	0.50
CD at 5%	-	-	-	1.74	1.56	1.25	1.36	1.96	1.39	1.39	2.48	1.42

Table 2 (b). Interaction effect of dates of sowing and fertilizer doses on plant height (cm) in kasuri methi

Treatments	**Plant height (cm)**											
	30 DAS			**60 DAS**			**90 DAS**			**At last harvest**		
	2016-17	**2017-18**	**Pooled**	**2016-17**	**2017-18**	**Pooled**	**2016-17**	**2017-18**	**Pooled**	**2016-17**	**2017-18**	**Pooled**
D_1F_1	7.27	7.44	7.35	25.30	26.56	25.93	56.19	59.30	57.74	62.12	59.50	60.81
D_1F_2	7.43	7.82	7.63	26.97	27.36	27.16	57.16	59.75	58.46	62.86	60.21	61.54
D_1F_3	7.62	7.99	7.81	26.49	26.93	26.71	56.55	59.49	58.02	61.89	59.70	60.80
D_1F_4	7.75	8.04	7.89	27.61	28.16	27.89	57.25	60.37	58.81	63.21	60.79	62.00
D_2F_1	7.12	7.27	7.20	25.92	25.83	25.87	55.74	58.02	56.88	59.59	58.90	59.25
D_2F_2	7.22	7.32	7.27	26.07	26.29	26.18	57.08	59.27	58.18	62.28	59.69	60.98
D_2F_3	7.32	7.54	7.43	25.85	26.00	25.93	56.15	58.79	57.47	61.45	59.08	60.27

D_2F_4	7.38	7.76	7.57	26.46	26.38	26.42	57.25	59.77	58.51	62.09	60.28	61.19
D_3F_1	6.79	6.71	6.75	20.57	22.77	21.67	53.12	51.07	52.10	56.50	57.30	56.90
D_3F_2	6.83	6.85	6.84	24.30	24.65	24.47	54.45	52.86	53.66	57.85	57.97	57.91
D_3F_3	6.86	7.03	6.95	22.73	23.97	23.35	53.53	52.34	52.93	57.51	57.87	57.69
D_3F_4	7.00	7.09	7.05	24.73	24.82	24.78	54.80	53.43	54.12	58.61	58.42	58.52
D_4F_1	6.35	6.19	6.27	22.33	24.26	23.30	51.28	46.40	48.84	53.06	53.96	54.40
D_4F_2	6.39	6.29	6.34	21.45	23.77	22.61	52.33	48.18	50.26	53.85	56.97	55.41
D_4F_3	6.61	6.43	6.52	20.81	24.04	22.43	51.69	47.31	49.50	53.61	55.98	54.80
D_4F_4	6.72	6.53	6.63	21.83	23.95	22.89	52.75	48.61	50.68	54.63	57.13	55.88
D_5F_1	5.93	5.43	5.68	17.64	18.86	18.25	44.62	40.73	42.68	44.09	55.75	51.46
D_5F_2	6.00	5.49	5.75	20.64	21.63	21.13	49.83	44.46	47.15	49.96	54.72	52.34
D_5F_3	6.11	5.55	5.83	17.79	19.42	18.61	49.25	40.90	45.08	49.29	53.19	51.82
D_5F_4	6.21	6.09	6.15	19.65	22.06	20.86	50.87	45.31	48.09	50.18	55.70	52.94
D_6F_1	5.21	4.92	5.07	19.28	20.57	19.92	47.93	43.13	45.53	48.96	52.02	48.06
D_6F_2	5.37	5.05	5.21	18.01	19.88	18.95	46.27	41.89	44.08	45.93	53.31	49.62
D_6F_3	5.45	5.23	5.34	19.47	21.11	20.29	45.53	43.59	44.56	45.40	54.35	49.30
D_6F_4	5.76	5.37	5.57	18.69	20.12	19.41	47.16	43.14	45.15	47.47	53.61	50.54
'F' test	NS	NS	NS	Sig.	Sig.	Sig.	Sig.	Sig.	Sig.	Sig.	Sig.	Sig.
SE(m) ±	0.14	0.13	0.10	0.62	0.55	0.44	0.48	0.69	0.49	0.49	0.87	0.50
CD at 5%	-	-	-	1.74	1.56	1.25	1.36	1.96	1.39	1.39	2.48	1.42

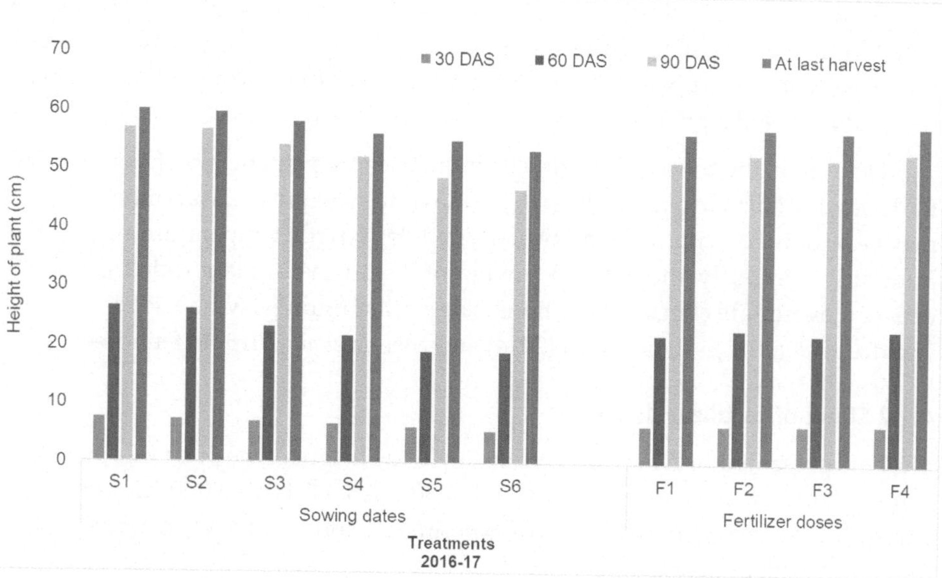

Fig. 2. Effect of dates of sowing and fertilizer doses on plant height in kasuri methi (cm) during 2016-17

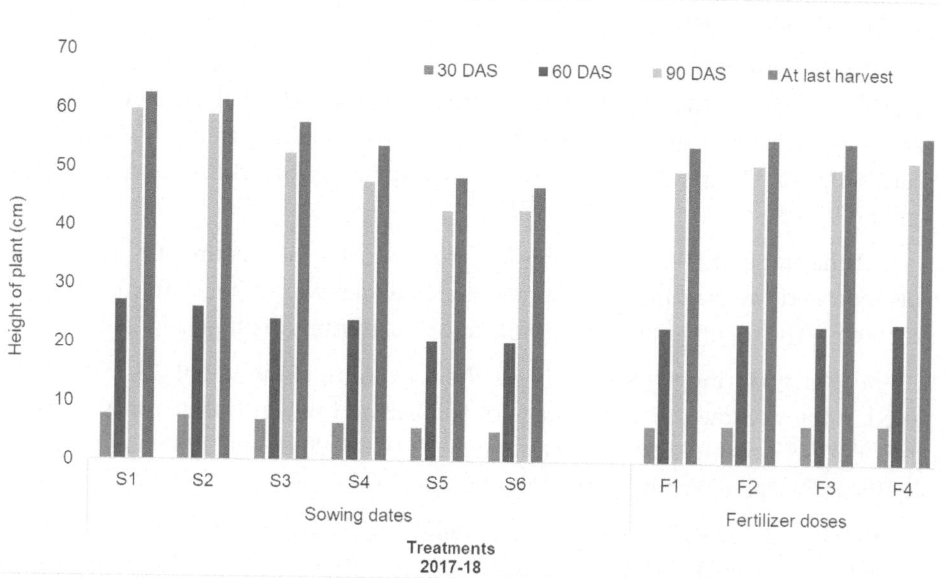

Fig. 3. Effect of dates of sowing and fertilizer doses on plant height in kasuri methi (cm) during 2017-18

which was at par with the treatment D_2 (59.49 cm). Whereas, the minimum plant height (53.32 cm) of kasuri methi was recorded in the treatment D_6.

In the year 2017-18, significantly the maximum plant height (62.52 cm) was noted in the sowing date 10th October (D_1) at the stage of last harvest and which was at par with the treatment D_2 (61.35 cm). Whereas, the sowing date 20th January (D_6) had recorded significantly the minimum plant height (46.94 cm).

Delayed in the sowing time directly influences the plant height of kasuri methi. Early sowing (10th October) reflects in the maximum height of kasuri methi plant. It may be attributed to the availability of desired day length, temperature and relative humidity. Contrarily, the delayed sowing of kasuri methi plant didn't get above effect of favourable climatic conditions. The similar findings were also reported by Nandre *et al.* (2011), Abou-Shleel (2014) and Meena *et al.* (2018) in fenugreek.

4.1.2 Effect of fertilizer doses

The data presented in Table 2 (a) revealed that, the height of the plant was increased subsequently at all the growth stages i.e. 30 DAS, 60 DAS, 90 DAS and at last harvesting stage during both years of experimentations, which were influenced by different fertilizer doses.

During the year 2016-17, the maximum height of kasuri methi plant was recorded in the treatments F_4 (6.80 cm) at the stage of 30 DAS and which was at par with treatment F_3 (6.66 cm). The treatment F_1 (6.44 cm) was recorded minimum plant height at this stage.

The maximum plant height (6.82 cm) was noted in fertilizer dose 50:50:25 kg/ha NPK (F_4) during the year 2017-18, at the stage of 30 DAS and which was significantly superior over rest of the treatments. However, significantly the minimum plant height (6.32 cm) was produced wherein application of fertilizer dose @ 40:40:20 kg/ha NPK (F_1) to the kasuri methi plant.

At the stage of 60 DAS, during the year 2016-17, the treatment F_4 (23.16 cm) was recorded the maximum plant height and which was at par with treatment F_2 (22.91 cm). The treatment F_1 gave significantly the minimum plant height (21.84 cm).

Similarly, during the year 2017-18, the maximum plant height (24.25 cm) was noted in the treatment F_4 at the stage of 60 DAS and which was at par with treatment F_2 (23.93 cm). Whereas, the treatment F_1 (40:40:20 kg/ha NPK) had recorded significantly the minimum plant height (23.14 cm).

In the year 2016-17, maximum plant height was recorded in treatment F_4 (53.35 cm) at the stage of 90 DAS and which was at par with treatment F_2 (52.86 cm). However, minimum plant height was noted in the treatment F_1 (51.48 cm).

At the stage of 90 DAS, during the year 2017-18, maximum plant height was registered (51.77 cm) in the treatment F_4 (50:50:25 kg/ha NPK) and which was at par with the treatment F_2 (51.07 cm). Whereas, minimum plant height (49.78 cm) was recorded in the treatment F_1 (40:40:20 kg/ha NPK).

At the last harvesting stage, plant height in the year 2016-17 was recorded the maximum (57.66 cm) by kasuri methi, which was fertilized with the 50:50:25 kg/ha NPK (F_4) and which was at par with treatment F_2 (57.14 cm). Whereas, significantly the minimum plant height (56.24 cm) was recorded with 40:40:20 kg/ha NPK (F_1).

Similarly, during the year 2017-18, the kasuri methi plant which was fertilized with F_4 attained maximum plant height (56.03 cm) and which was at par with F_2 (55.46 cm). However, minimum plant height (54.05 cm) was recorded under the treatment F_1 (40:40:20 kg/ha NPK).

The kasuri methi plant fertilized with higher dose exhibit more plant height as against rest of the levels of the fertilizer. The highest fertilizer level 50:50:25 kg ha^{-1} NPK (F_4) expressed maximum plant height, might be due to fact that, nitrogen is required for cell elongation, phosphorus would be responsible for nitrogenous activity in root zone and potassic fertilizer improves the reproductive growth i.e. overall vegetative growth in respect of plant height would be possible only because of application of highest level of NPK in kasuri methi. The results are in agreement with the findings of Kumar *et al.* (2015), Meena *et al.* (2018) in fenugreek, Anupama (2012) and Sunanda *et al.* (2014) in kasuri methi.

4.1.3 Interaction effect

The data presented in Table 2 (b) indicated that, the interaction effect between sowing dates and fertilizer doses on plant height were found to be non-significant at 30 DAS. Whereas, at 60, 90 and at last harvest these were found to be significant.

The periodical growth observations on height of plant were found to be increased at faster rate up to 90 DAS in all the treatments and which were increased slowly up to the last harvesting stage of the crop.

During the experimental year 2016-17, significantly the maximum plant height was recorded at 60 DAS, 90 DAS and at last harvest by the treatment combination of D_1F_4 (27.61, 57.25 and 63.21 cm, respectively) as compared to all other treatment combinations. However, significantly the minimum plant height was recorded at 60 DAS, 90 DAS and at last harvest by the treatment combination D_6F_1 (19.28, 47.93 and 48.96 cm, respectively).

In 2017-18, significantly the maximum plant height was observed at 60 DAS, 90 DAS and at last harvest by the treatment combination D_1F_4 (28.16, 60.37 and 60.79 cm, respectively) as compared to all other treatments. However, significantly the minimum plant height was observed at 60 DAS, 90 DAS and at last harvest by the treatment combination D_6F_1 (20.57, 43.13 and 52.02 cm, respectively).

The interaction effect between sowing dates and fertilizer doses indicated the maximum plant height in earlier (10th October) sowing along with the maximum (50:50:25 kg ha^{-1} NPK) fertilization in the present study. This might be due to getting the quick start for vegetative growth and simultaneously supplied with all primary nutrients in desired quantity compared to rest of the treatment combinations. The results of the present investigation are supported by the findings of Ali *et al.* (2015) in coriander.

4.2 BRANCHES PER PLANT

4.2.1 Effect of sowing dates

The data presented in Table 3 (a) revealed that, at all the stages of observations, the number of branches per plant were significantly influenced by the sowing time.

In the year 2016-17, significantly the maximum number of branches per plant were recorded in the sowing date 10th October (D_1) at 30 DAS (2.00), 60 DAS (5.05), 90 DAS (7.59) and at the last harvesting stage (8.46). However, significantly the minimum number of branches per plant were recorded in the sowing date 20th January (D_6) at 30 DAS (1.00), 60 DAS (3.37), 90 DAS (4.61) and last harvesting stage (6.69).

At the stages of 30 DAS, 60 DAS, 90 DAS and at the last harvesting stage, during the year 2017-18, significantly the maximum number of branches (1.99, 5.63, 7.88 and 8.47, respectively) were exhibited with the sowing date 10th October (D_1). However, significantly the minimum number of branches per plant were recorded in the sowing date 20th January (D_6) at the stage of 30 DAS (0.79), 60 DAS (3.05), 90 DAS (5.48) and the last harvesting stage (6.45). The better vegetative growth of kasuri methi crop might be achieved due to an early sowing (10th October) and congenial climatic conditions availed by the crop during the growth period. The similar findings were also reported by Nandre *et al.* (2011), Sowmya (2016) and Meena *et al.* (2018) in fenugreek.

4.2.2 Effect of fertilizer doses

From the data presented in Table 3 (a), it is found that, at all the stages of observations, branches per plant were significantly influenced by the fertilizer doses.

During the year 2016-17, at 30 DAS, 60 DAS, 90 DAS and at last harvesting stage, the maximum number of branches per plant (1.63, 4.49, 6.39 and 7.83, respectively) were observed in the fertilizer dose 50:50:25

Table 3 (a). Effect of dates of sowing and fertilizer doses on number of branches per plant in kasuri methi

Treatments	Number of branches											
	30 DAS			60 DAS			90 DAS			At last harvest		
	2016-17	2017-18	Pooled	2016-17	2017-18	Pooled	2016-17	2017-18	Pooled	2016-17	2017-18	Pooled
Dates of sowing (D)												
D_1 - 10th October	2.00	1.99	2.00	5.05	5.63	5.34	7.59	7.88	7.73	8.46	8.47	8.47
D_2 - 30th October	1.90	1.89	1.89	4.20	5.51	5.21	7.34	7.69	7.51	8.30	8.28	8.29
D_3 - 20th November	1.59	1.77	1.68	4.68	5.06	4.87	6.59	7.26	6.93	7.92	7.90	7.91
D_4 - 10th December	1.44	1.45	1.44	4.41	4.54	4.48	5.87	6.64	6.26	7.62	7.50	7.56
D_5 - 30th December	1.12	1.17	1.15	3.81	3.73	3.77	4.96	5.74	5.36	6.96	6.69	6.82
D_6 - 20th January	1.00	0.79	0.89	3.37	3.05	3.21	4.61	5.48	5.05	6.69	6.45	6.57
'F' test	Sig.	Sig.	Sig.	Sig.	Sig.	Sig.	Sig.	Sig.	Sig.	Sig.	Sig.	Sig.
SE(m) ±	0.07	0.04	0.04	0.05	0.07	0.04	0.09	0.16	0.10	0.07	0.07	0.04
CD at 5%	0.19	0.11	0.11	0.14	0.20	0.13	0.26	0.45	0.27	0.19	0.20	0.11
Fertilizer dose (F)												
F_1 - 40:40:20 kg ha^{-1} NPK	1.40	1.43	1.42	4.22	4.38	4.31	5.87	6.58	6.22	7.46	7.37	7.14
F_2 - 40:50:25 kg ha^{-1} NPK	1.47	1.48	1.48	4.42	4.68	4.55	6.33	6.80	6.57	7.74	7.62	7.68
F_3 - 50:40:20 kg ha^{-1} NPK	1.53	1.52	1.53	4.35	4.56	4.45	6.05	6.68	6.36	7.60	7.49	7.55
F_4 - 50:50:25 kg ha^{-1} NPK	1.63	1.61	1.62	4.49	4.73	4.61	6.39	7.07	6.73	7.83	7.72	7.77

'F' test	Sig.	Sig.	Sig.	Sig.	Sig.	Sig.	Sig.	Sig.	Sig.	Sig.	Sig.	Sig.
SE(m) ±	0.05	0.03	0.03	0.04	0.06	0.04	0.08	0.13	0.08	0.05	0.06	0.03
CD at 5%	0.15	0.09	0.09	0.12	0.16	0.10	0.21	0.37	0.22	0.15	0.16	0.09
Interaction (D X F)												
'F' test	NS	NS	NS	NS	NS	NS	Sig.	Sig.	Sig.	Sig.	Sig.	Sig.
SE(m) ±	0.13	0.08	0.08	0.10	0.14	0.09	0.18	0.31	0.19	0.13	0.14	0.08
CD at 5%	-	-	-	-	-	-	0.52	0.90	0.55	0.37	0.39	0.23

Table 3 (b). Interaction effect of dates of sowing and fertilizer doses on number of branches per plant in kasuri methi

Treat-ments	Number of branches											
	30 DAS			60 DAS			90 DAS			At last harvest		
	2016-17	2017-18	Pooled	2016-17	2017-18	Pooled	2016-17	2017-18	Pooled	2016-17	2017-18	Pooled
D_1F_1	1.95	1.89	1.92	4.92	5.47	5.20	7.39	7.71	7.55	8.26	8.27	8.27
D_1F_2	1.99	2.00	2.00	5.13	5.70	5.42	7.80	7.92	7.86	8.53	8.54	8.54
D_1F_3	2.01	2.01	2.01	5.00	5.63	5.32	7.31	7.79	7.55	8.46	8.47	8.47
D_1F_4	2.06	2.06	2.06	5.14	5.72	5.43	7.86	8.10	7.98	8.59	8.60	8.60
D_2F_1	1.74	1.87	1.80	4.84	5.39	5.12	6.91	7.52	7.22	8.17	8.17	8.17
D_2F_2	1.90	1.73	1.82	4.92	5.56	5.24	7.60	7.73	7.66	8.40	8.30	8.35
D_2F_3	1.87	1.95	1.91	4.89	5.47	5.18	7.36	7.66	7.51	8.21	8.24	8.23
D_2F_4	2.08	2.00	2.04	4.97	5.61	5.29	7.48	7.85	7.66	8.43	8.40	8.42

D_3F_1	1.59	1.69	1.64	4.56	4.89	4.72	6.31	7.09	6.70	7.72	7.72	7.72
D_3F_2	1.44	1.76	1.60	4.74	5.16	4.95	6.74	7.32	7.03	7.96	8.00	7.98
D_3F_3	1.51	1.80	1.65	4.66	4.97	4.82	6.51	7.22	6.86	7.89	7.81	7.85
D_3F_4	1.81	1.83	1.82	4.77	5.22	4.99	6.79	7.42	7.11	8.10	8.08	8.09
D_4F_1	1.38	1.39	1.39	4.30	4.41	4.35	5.59	6.28	5.94	7.44	7.31	7.38
D_4F_2	1.31	1.41	1.36	4.45	4.56	4.51	5.98	6.73	6.36	7.70	7.56	7.63
D_4F_3	1.41	1.47	1.44	4.41	4.50	4.46	5.81	6.58	6.20	7.54	7.46	7.50
D_4F_4	1.64	1.52	1.58	4.50	4.68	4.59	6.11	6.98	6.55	7.78	7.66	7.72
D_5F_1	1.10	1.07	1.08	3.63	3.48	3.55	4.09	4.57	4.33	6.29	6.03	6.16
D_5F_2	1.05	1.13	1.09	3.86	3.87	3.86	5.33	6.02	5.67	7.18	6.93	7.06
D_5F_3	1.14	1.23	1.19	3.75	3.66	3.71	5.02	6.06	5.54	7.06	6.78	6.92
D_5F_4	1.20	1.27	1.24	3.99	3.93	3.96	5.42	6.34	5.88	7.29	7.01	7.15
D_6F_1	0.67	0.69	0.68	3.10	2.67	2.88	4.94	6.29	5.62	6.89	6.68	6.79
D_6F_2	1.12	0.76	0.94	3.43	3.20	3.32	4.52	5.12	4.82	6.65	6.36	6.51
D_6F_3	1.25	0.71	0.98	3.37	3.10	3.24	4.29	4.74	4.52	6.46	6.19	6.33
D_6F_4	0.96	0.98	0.97	3.59	3.24	3.42	4.70	5.76	5.23	6.77	6.56	6.67
'F' test	NS	NS	NS	NS	NS	NS	Sig.	Sig.	Sig.	Sig.	Sig.	Sig.
SE(m) ±	0.13	0.08	0.08	0.10	0.14	0.09	0.18	0.31	0.19	0.13	0.14	0.08
CD at 5%	-	-	-	-	-	-	0.52	0.90	0.55	0.37	0.39	0.23

kg/ha NPK (F_4). However, significantly the minimum branches per plant were registered in the fertilizer dose 40:40:20 kg/ha NPK (F_1), at the stages of 30 DAS (1.40), 60 DAS (4.22), 90 DAS (5.87) and at last harvesting stage (7.46).

In the year 2017-18, maximum branches per plant were produced in the treatment F_4, at the stages of 30 DAS (1.61), 60 DAS (4.73), 90 DAS (7.07) and at the last harvesting stage (7.72). Whereas, at the stages of 30 DAS, 60 DAS, 90 DAS and at the last harvesting stage, minimum branches per plant (1.43, 4.38, 6.58 and 7.37, respectively) were recorded in the treatment F_1 (40:40:20 kg/ha NPK).

This might be due to the fact that, availability of primary nutrients through inorganic sources might have enhanced the meristematic activities and size of cell and formation and functioning of protoplasm, which consequently improved the number of branches per plant. Similarly, the maximum height of plant might have produced the maximum number of branches per plant. The similar results were obtained by Deora *et al.* (2009), Sunanda *et al.* (2014) in kasuri methi, Kumar *et al.* (2015) and Meena *et al.* (2018) in fenugreek.

4.2.3 Interaction effect

The data presented in Table 3 (b) revealed that, the interaction effects due to the sowing dates and fertilizer doses on branches per plant were found to be non-significant at the growth stages of 30 DAS and 60 DAS. While, those were observed to be significant at 90 DAS and at the last harvesting stage of crop, during both the years of experimentations (2016-17 and 2017-18).

During the experimental year 2016-17, significantly the maximum number of branches were recorded at 90 DAS and at last harvest by the treatment combination D_1F_4 (7.86 and 8.59, respectively) as compare to all other treatment combinations. However, significantly the minimum branches per plant were recorded at 90 DAS and at last harvest by the treatment combination of D_6F_3 (4.29 and 6.46, respectively).

In 2017-18, significantly the maximum number of branches were observed at 90 DAS and at last harvest by the treatment combination of D_1F_4 (8.10 and 8.60, respectively) as compared to all other treatments. However, significantly the minimum number of branches were observed at 90 DAS and at last harvest by the treatment combination of D_6F_3 (4.74 and 6.19, respectively) in kasuri methi.

Kasuri methi crop sown on 10th October (D_1) and fertilized with 50:50:25 kg ha^{-1} NPK produced the maximum number of branches per plant, because of availability of ample quantity of primary nutrients through inorganic fertilizers with explicitly influence of environment (early sowing) on growth, development and yield of kasuri methi. Bhutia and Sharangi (2016) were also reported the results in similar lines in fenugreek in kasuri methi.

4.3 LEAVES PER PLANT

The data presented in Table 4 (a) and (b) revealed that, at all the stages of observation, the number of leaves per plant were significantly influenced by the sowing dates and fertilizer levels and depicted in Fig. 4 and 5.

4.3.1 Effect of sowing dates

The data presented in Table 4 (a) revealed that, at all the stages of observations, the number of leaves per plant were significantly influenced by the sowing dates.

In the year 2016-17, significantly the maximum number of leaves per plant were counted in the treatment D_1 at 30 DAS (11.12), 60 DAS (25.00), 90 DAS (104.84) and at the time of last harvest (118.70). However, significantly the minimum number of leaves per plant were recorded in the sowing date 20th January (D_6) at 30 (08.09), 60 (18.25), 90 (57.03) and at the time of last harvest (87.46).

At the stages of 30 DAS, 60 DAS, 90 DAS and at the time of last harvest, during the year 2017-18 significantly the maximum number of leaves (11.28, 25.60, 109.54 and 121.93, respectively) were exhibited with

Table 4 (a). Effect of dates of sowing and fertilizer doses on number of leaves in kasuri methi

Treatments	Number of leaves											
	30 DAS			60 DAS			90 DAS			At last harvest		
	2016-17	2017-18	Pooled	2016-17	2017-18	Pooled	2016-17	2017-18	Pooled	2016-17	2017-18	Pooled
Dates of sowing (D)												
D_1 - 10th October	11.12	11.28	11.20	25.00	25.60	25.30	104.84	109.54	107.19	118.70	121.93	120.32
D_2 - 30th October	10.80	11.04	11.92	24.04	25.07	24.56	102.87	107.28	105.07	117.07	119.00	118.04
D_3 - 20th November	10.11	10.45	10.28	22.74	23.89	23.31	93.14	94.13	93.63	107.05	111.04	109.04
D_4 - 10th December	9.53	9.89	9.71	21.29	23.00	22.14	80.06	75.22	77.64	100.69	99.99	100.34
D_5 - 30th December	8.60	9.43	9.02	18.83	20.91	19.87	61.86	59.27	60.56	90.91	91.22	91.07
D_6 - 20th January	8.09	8.81	8.45	18.25	20.16	19.21	57.03	53.58	53.30	87.46	86.65	87.05
'F' test	Sig.	Sig.	Sig.	Sig.	Sig.	Sig.	Sig.	Sig.	Sig.	Sig.	Sig.	Sig.
SE(m) ±	0.15	0.09	0.08	0.09	0.22	0.11	0.92	0.85	0.63	0.63	0.45	0.36
CD at 5%	0.42	0.24	0.24	0.25	0.63	0.30	2.62	2.42	1.79	1.80	1.27	1.02
Fertilizer dose (F)												
F_1 - 40:40:20 kg ha^{-1} NPK	9.44	9.92	9.68	21.25	22.67	21.96	80.95	79.79	80.37	101.14	102.50	101.83
F_2 - 40:50:25 kg ha^{-1} NPK	9.66	10.09	9.87	21.76	23.34	22.55	83.69	84.23	83.96	104.43	105.90	105.16
F_3 - 50:40:20 kg ha^{-1} NPK	9.67	10.24	9.95	21.68	22.93	22.31	83.07	82.67	82.87	102.75	104.10	103.44

F_4 - 50:50:25 kg ha^{-1} NPK	10.06	10.36	10.21	22.08	23.48	22.78	85.49	85.99	85.74	106.26	107.40	106.81
'F' test	Sig.	Sig.	Sig.	Sig.	Sig.	Sig.	Sig.	Sig.	Sig.	Sig.	Sig.	Sig.
SE(m) ±	0.12	0.07	0.07	0.09	0.18	0.09	0.75	0.69	0.51	0.51	0.36	0.29
CD at 5%	0.34	0.20	0.20	0.25	0.52	0.25	2.14	1.98	1.46	1.47	1.04	0.83
Interaction (D X F)												
'F' test	NS	NS	NS	Sig.	Sig.	Sig.	Sig.	Sig.	Sig.	Sig.	Sig.	Sig.
SE(m) ±	0.30	0.17	0.17	0.17	0.44	0.21	1.84	1.70	1.26	1.27	0.89	0.71
CD at 5%	-	-	-	0.50	1.26	0.61	5.24	4.84	3.58	3.61	2.54	2.03

Table 4 (b). Interaction effect of dates of sowing and fertilizer doses on number of leaves per plant in kasuri methi

Treatments	Number of leaves											
	30 DAS			60 DAS			90 DAS			At last harvest		
	2016-17	2017-18	Pooled	2016-17	2017-18	Pooled	2016-17	2017-18	Pooled	2016-17	2017-18	Pooled
D_1F_1	10.57	11.01	10.79	24.30	25.08	24.69	103.16	106.27	104.72	115.93	119.76	117.85
D_1F_2	10.98	11.27	11.12	25.21	25.85	25.53	105.38	110.89	108.14	119.84	122.87	121.36
D_1F_3	11.42	11.35	11.38	24.98	25.43	25.21	104.40	108.94	106.67	118.30	121.31	119.81
D_1F_4	11.52	11.48	11.50	25.50	26.04	25.77	106.42	112.08	109.25	120.72	123.79	122.25
D_2F_1	10.44	10.80	10.62	23.28	24.86	24.07	100.83	105.65	103.24	115.19	116.61	115.90
D_2F_2	10.71	10.95	10.83	24.22	25.10	24.66	103.68	107.96	105.82	117.78	119.96	118.87

D_2F_3	10.97	11.12	11.05	23.83	24.89	24.36	101.73	107.23	104.48	116.41	118.24	117.32
D_2F_4	11.08	11.30	11.19	24.83	25.44	25.14	105.23	108.26	106.75	118.90	121.20	120.05
D_3F_1	9.95	10.28	10.11	22.27	22.36	22.32	75.57	90.91	83.24	104.83	108.32	106.57
D_3F_2	10.02	10.38	10.20	22.87	24.62	23.74	98.51	95.00	96.75	107.40	111.80	109.60
D_3F_3	10.15	10.54	10.34	22.64	23.99	23.32	97.68	93.63	95.65	105.28	110.42	107.85
D_3F_4	10.30	10.61	10.45	23.16	24.57	23.87	100.79	96.98	98.89	110.69	113.62	112.16
D_4F_1	9.27	9.72	9.50	21.32	23.68	22.50	94.01	69.45	81.73	96.81	97.83	97.32
D_4F_2	9.49	9.83	9.66	20.99	23.07	22.03	73.28	77.24	75.26	101.10	100.84	100.97
D_4F_3	9.61	9.96	9.79	21.68	22.63	22.16	77.43	74.49	75.96	99.62	98.59	99.11
D_4F_4	9.75	10.05	9.90	21.18	22.60	21.89	75.52	79.71	77.62	105.21	102.68	103.95
D_5F_1	8.63	9.25	8.94	17.29	19.31	18.30	53.84	48.36	51.10	83.53	83.38	83.46
D_5F_2	8.84	9.36	9.10	19.31	21.46	20.38	64.66	62.48	63.57	93.56	94.33	93.95
D_5F_3	7.68	9.51	8.60	19.18	20.98	20.08	61.95	60.83	61.39	92.58	91.81	92.20
D_5F_4	9.24	9.61	9.43	19.55	21.91	20.73	66.98	65.39	66.19	93.96	95.39	94.68
D_6F_1	7.79	8.47	8.13	19.02	20.73	19.88	58.26	58.09	58.17	90.52	89.23	89.88
D_6F_2	7.93	8.72	8.33	17.97	19.95	18.96	56.64	51.79	54.22	86.93	85.45	86.19
D_6F_3	8.18	8.93	8.55	17.77	19.65	18.71	55.22	50.89	53.05	84.33	84.37	84.35
D_6F_4	8.45	9.11	8.78	18.25	20.33	19.29	58.02	53.53	55.77	88.05	87.52	87.79
'F' test	NS	NS	NS	Sig.	Sig.	Sig.	Sig.	Sig.	Sig.	Sig	Sig.	Sig.
SE(m) ±	0.30	0.17	0.17	0.17	0.44	0.21	1.84	1.70	1.26	1.27	0.89	0.71
CD at 5%	-	-	-	0.50	1.26	0.61	5.24	4.84	3.58	3.61	2.54	2.03

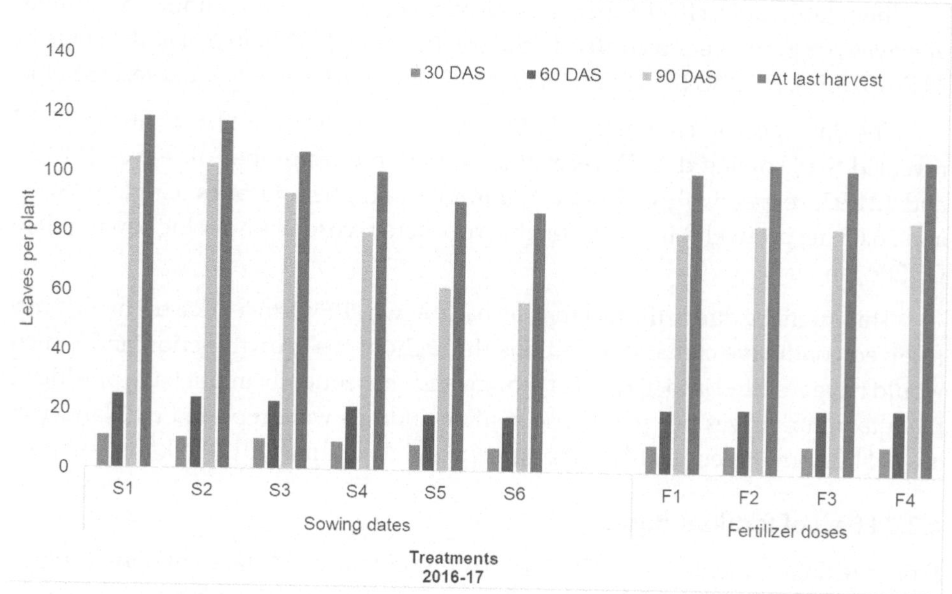

Fig. 4. Effect of dates of sowing and fertilizer doses on number of leaves in kasuri methi during 2016-17

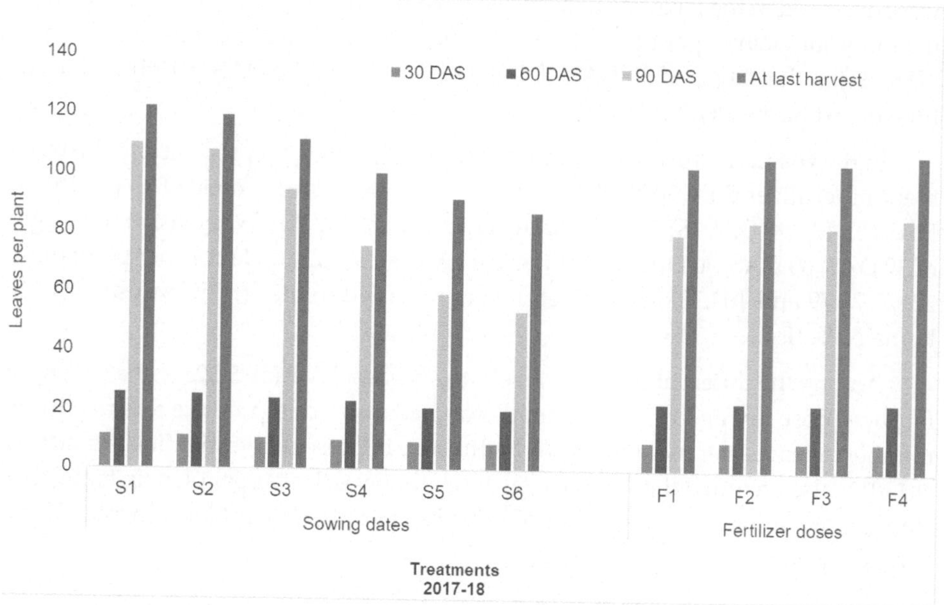

Fig. 5 Effect of dates of sowing and fertilizer doses on number of leaves in kasuri methi during 2017-18

the treatment D_1 (10th October). However, significantly the minimum number of leaves per plant were recorded in the treatment D_6 (20th January) at the stage of 30 DAS (8.81), 60 DAS (20.16), 90 DAS (53.83) and the time of last harvest (86.65).

The data recorded at the last harvesting stage in the year 2016-17 and 2017-18 revealed that, sowing date D_1 recorded the maximum number of leaves (118.70 and 121.93, respectively). However, minimum number of leaves per plant (87.05 and 86.65, respectively) in kasuri methi crop were recorded under the sowing date D_6 (20th January).

This might be due to the fact that, an early sown (10th October) kasuri methi crop received conducive climatic conditions throughout the growth period and which would have resulted into luxurious growth and consequently might have produced the maximum leaves per plant. The similar findings were reported by Nandre *et al.* (2011), Abou-Shleel (2014) in fenugreek, and Anupama (2012) in kasuri methi.

4.3.2 Effect of fertilizer doses

From the data presented in Table 4 (a), it is found that, at all stages of observations, leaves per plant were significantly influenced by the fertilizer doses.

During the year 2016-17, at the stages of 30 DAS, 60 DAS, 90 DAS and at the last harvest, the maximum leaves per plant (10.06, 22.08, 85.49 and 106.26, respectively) were observed in the fertilizer dose 50:50:25 kg/ha NPK (F_4). However, significantly the minimum leaves per plant were registered in the fertilizer dose 40:40:20 kg/ha NPK (F_1) at the stages of 30 DAS (9.44), 60 DAS (21.25), 90 DAS (80.95) and at the time of last harvest (101.14).

In the year 2017-18, the maximum leaves per plant were produced in the treatment of fertilizer dose 50:50:25 kg/ha NPK (F_4), at the stages of 30 DAS (10.36), 60 DAS (23.48), 90 DAS (85.99) and at the last harvest (107.40). Whereas, at the stages of 30 DAS, 60 DAS, 90 DAS and at the last harvest, minimum leaves per plant (9.92, 22.67, 79.79 and 102.50, respectively) were recorded in the fertilizer dose 40:40:20 kg/ha NPK (F_1).

Among the different fertilizer levels, the highest level (50:50:25 kg ha^{-1} NPK) of fertilizer dose applied to kasuri methi crop had noticed to produce the maximum number of leaves per plant in kasuri methi crop in present investigation. Nandre *et al.* (2011) also recorded the results regarding number of leaves per plant in fenugreek stating that, an increasing number of branches per plant might have increased the leaves per plant.

4.3.3 Interaction effect

The data presented in Table 4 (b) revealed that, the interaction effect due to the sowing dates and fertilizer doses on per plant leaves were found to be non-significant at the growth stage of 30 DAS while, these were found to be significant at the stage of 60 DAS, 90 DAS and at the harvest, during both the years of experimentations (2016-17 and 2017-18).

During the experimental year 2016-17, significantly the maximum number of leaves were recorded at 60 DAS, 90 DAS and at last harvest by the treatment combination D_1F_4 (25.50, 106.42 and 120.72, respectively) as compared to all other treatment combinations. However, significantly the minimum leaves per plant were produced at 60 DAS, 90 DAS and at last harvest by the treatment combination of D_6F_3 (17.77, 55.22 and 84.33, respectively).

In 2017-18, significantly the maximum number of leaves were observed at 60 DAS, 90 DAS and at last harvest by the treatment combination of D_1F_4 (26.04, 112.08 and 123.79, respectively) as compared to all other treatments. However, significantly the minimum number of leaves per plant were noticed at 60 DAS, 90 DAS and at last harvest by the treatment combination D_6F_3 (19.65, 50.89 and 84.37, respectively).

The early sowing of kasuri methi crop along with higher level of fertilizers resulted in the maximum number of branches per plant. More the number of branches more would be the number of leaves in case of kasuri methi, as reported by the earlier workers Nandre *et al.* (2011) and Bhutia and Sharangi (2016). Further higher level of nitrogen, phosphorus and potassium generally produced more number of secondary branches per plant in fenugreek as reported by Datta *et al.* (2017), which is in line with the results obtained in the present investigation.

B. YIELD ATTRIBUTES AND YIELD

4.4 DAYS TO FLOWER INITIATION

The data in respect of days to flower initiation during the years 2016-17 and 2017-18 as influenced by different dates of sowing and fertilizer doses are presented in Table 5.

4.4.1 Effect of sowing dates

The data presented in Table 5 revealed that, the treatment differences towards days to flower initiation as influenced due to the sowing dates were found to be significant during both the years of experimentations.

In the year 2016-17, the sowing date 30th October (D_2) was recorded significantly the maximum period for flower initiation (53.67 days) and was followed by sowing date 10th October (D_1) which was 52.00 days. However, sowing date 20th January (D_6) was recorded significantly the minimum period for flower initiation (48.52 days) and this was found to be at par with the treatment D_5 (48.80 days).

Similarly, during 2017-18, sowing date 30th October (D_2) was recorded significantly the maximum period for flower initiation (53.96 days) and was followed by sowing date 10th October (D_1) which was 52.91 days. The sowing date 20th January

(D_6) was reported significantly the minimum period (47.06 days) for flower initiation in kasuri methi crop.

It appears that, sowing of kasuri methi seed on 30th October (D_2) delayed the flower initiation by 4-5 days as compared to sowing of 20th January (D_6). The wide range was observed for getting the flowering in kasuri methi crop due to different sowing dates might be ascribed due to weather variation particularly temperature differences as first sowing date and at last sowing date. The similar findings were reported by Nandre *et al.* (2011) in fenugreek.

4.4.2 Effect of fertilizer doses

Days required for flower initiation as influenced due to the sowing dates were found to be significant during both the years of experimentations which is furnished in Table 5.

Table 5. Effect of dates of sowing and fertilizer doses on days to flower initiation in kasuri methi

Treatments	Days to flower initiation		
	2016-17	2017-18	Pooled
Dates of sowing (D)			
D_1 - 10th October	52.00	52.51	52.25
D_2 - 30th October	53.67	53.96	53.81
D_3 - 20th November	50.58	50.71	50.69
D_4 - 10th December	49.66	48.84	49.25
D_5 - 30th December	48.80	47.42	48.11
D_6 - 20th January	48.52	47.06	47.49
'F' test	Sig.	Sig.	Sig.
SE(m) ±	0.10	0.13	0.10
CD at 5%	0.28	0.38	0.27
Fertilizer dose (F)			
F_1 - 40:40:20 kg ha^{-1} NPK	51.02	50.76	50.89
F_2 - 40:50:25 kg ha^{-1} NPK	50.31	49.73	50.02
F_3 - 50:40:20 kg ha^{-1} NPK	50.72	50.39	50.55
F_4 - 50:50:25 kg ha^{-1} NPK	50.11	49.52	49.82

'F' test	Sig.	Sig.	Sig.
SE(m) ±	0.08	0.11	0.08
CD at 5%	0.23	0.31	0.22
Interaction (D X F)			
'F' test	NS	NS	NS
SE(m) ±	0.20	0.27	0.19
CD at 5%	-	-	-

The fertilizer dose 40:40:20 kg/ha NPK (F_1) reported maximum days to flower initiation (51.02 days) during the year 2016-17, which was found to be significantly superior over rest of the treatments. However, fertilizer dose 50:50:25 kg/ha NPK (F_4) recorded significantly the minimum days to flower initiation (50.11 days) and which was found to at par with the treatment F_2 (40:50:25 kg/ha NPK).

In subsequent year 2017-18, treatment F_1 (40:40:20 kg/ha NPK) took the maximum days for flower initiation (50.76 days). While, minimum days were required for flower initiation (49.52 days) in treatment F_4 (50:50:25 kg/ha NPK) and which was found to be at par with the treatment F_2 (40:50:25 kg/ha NPK). The delay in flowering in kasuri methi supplied with different levels of primary nutrients might be due to the fact that, lower levels of nutrients delayed the flowering mainly because of poor growth parameters which might have passed through long span and ultimately resulted in delayed flowering. The results of the present investigation are supported by the findings of Datta *et al.* (2017) in fenugreek.

4.4.3 Interaction effect

All interaction effect of dates of sowing and fertilizer doses on the days required for flower initiation independently during both the years of experimentations 2016-17 and 2017-18 were found to be non-significant.

4.5 DAYS TO 50 PER CENT FLOWERING

The data in respect of days required for 50 per cent flowering during the year 2016-17 and 2017-18 as influenced by sowing dates and fertilizer doses were presented in Table 6 and illustrated in Fig. 6.

4.5.1 Effect of sowing dates

The data presented in Table 6 revealed that, different sowing dates showed significant effect on days required for 50 per cent flowering during both the years of experimentations.

Table 6. Effect of dates of sowing and fertilizer doses on days to 50 per cent flowering in kasuri methi

Treatments	Days 50 per cent flowering		
	2016-17	2017-18	Pooled
Dates of sowing (D)			
D_1 - 10th October	58.55	57.51	58.03
D_2 - 30th October	60.09	58.75	59.42
D_3 - 20th November	57.49	56.24	56.86
D_4 - 10th December	56.31	55.12	55.71
D_5 - 30th December	54.58	54.35	54.46
D_6 - 20th January	54.35	53.84	53.09
'F' test	Sig.	Sig.	Sig.
SE(m) ±	0.09	0.20	0.12
CD at 5%	0.26	0.58	0.33
Fertilizer dose (F)			
F_1 - 40:40:20 kg ha^{-1} NPK	57.35	56.44	56.89
F_2 - 40:50:25 kg ha^{-1} NPK	56.69	55.79	56.24
F_3 - 50:40:20 kg ha^{-1} NPK	57.05	56.08	56.57
F_4 - 50:50:25 kg ha^{-1} NPK	56.49	55.56	56.03
'F' test	Sig.	Sig.	Sig.
SE(m) ±	0.08	0.17	0.10
CD at 5%	0.21	0.47	0.27
Interaction (D X F)			
'F' test	NS	NS	NS
SE(m) ±	0.18	0.41	0.23
CD at 5%	-	-	-

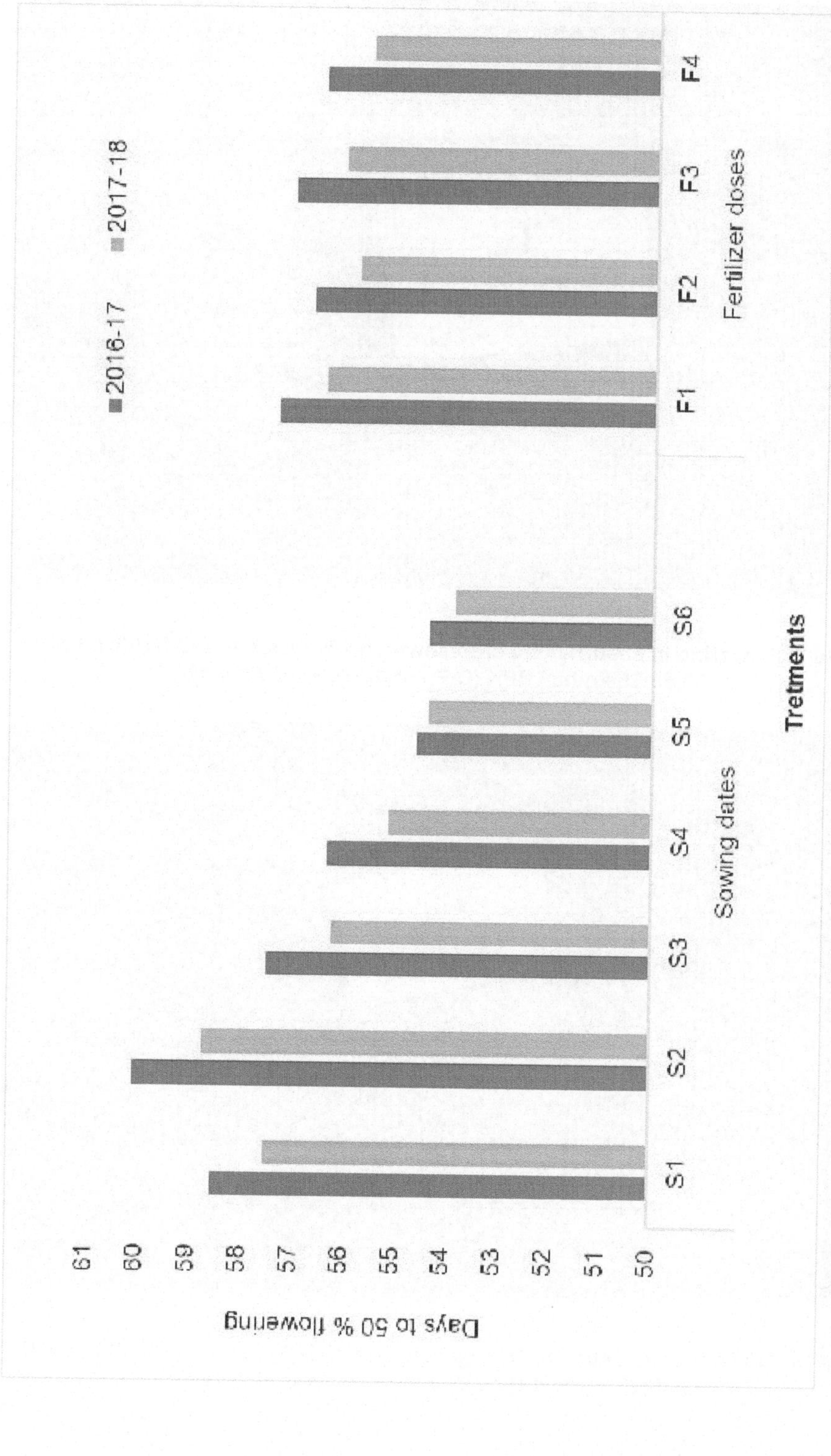

Fig. 6. Effect of dates of sowing and fertilizer doses on days to 50 per cent flowering in kasuri methi

Plate 3. Flowering in kasuri methi crop sown on 10th October and fertilized with 50:50:25 kg ha^{-1} NPK (D_1F_4) in the year of 2016-17

Plate 4. Flowering in kasuri methi crop sown on 20th January and fertilized with 40:40:20 kg ha-1 NPK (D6F1) in the year of 2016-17

In the year 2016-17, the maximum (60.09 days) days to 50 per cent flowering was observed in the sowing date 30th October (D_2) and was followed by sowing date 10th October (D_1) which was 58.55 days. While, sowing date 20th January (D_6) measured significantly the minimum (54.35 days) days to 50 per cent flowering and which was found to be at par with the treatment D_5 (54.58 days).

During the year 2017-18, sowing date 30th October (D_2) noted the maximum (58.75 days) days required for 50 per cent flowering and was followed by sowing date 10th October (D_1) which was 57.51 days. Whereas, the minimum (53.84 days) days required for 50 per cent flowering were noted in sowing date 20th January (D_6) and which was found to be at par with the treatment D_5 (54.35 days).

There was a trend of earlier flowering with advancement of sowing dates starting from 10th October upto 20th January; keeping 20 days interval, with the exception in case of 30th October sowing of kasuri methi crop in the present investigation. Naturally, crop grows in the congenial climatic conditions especially the desired temperature availability upto its completion of vegetative growth and its reflection into initiation of reproductive phase in the form of the more number of days require for 50 per cent flowering. The results of the present investigation are supported by the findings of Nandre *et al.* (2011) in kasuri methi.

4.5.2 Effect of fertilizer doses

The perusal of data from Table 6 revealed that, the different fertilizer doses on days required for 50 per cent flowering were found to be significant, during the years 2016-17 and 2017-18.

During the year 2016-17 and 2017-18, the fertilizer dose 40:40:20 kg/ha NPK (F_1) took the maximum days for 50 per cent flowering (57.35 and 56.44 days, respectively) and which was found to be significantly superior over rest of the treatments. However, fertilizer dose 50:50:25 kg/ha NPK (F_4) was recorded minimum days for 50 per cent flowering (56.49 and 55.56 days, respectively) and which was found to be at par with the treatment 40:50:25 kg/ha NPK (F_2).

The lower levels of nutrient application to kasuri methi, delayed the initiation as well as 50 per cent flowering in the present investigation. The probable reason for this might be the slow and prolonged availability of nutrients, which may have resulted in the delayed in flowering. This results are in line with Sunanda *et al.* (2014) in kasuri methi.

4.5.3 Interaction effect

Interaction effect of sowing dates and fertilizer doses on the days required for 50 per cent flowering, during both the years of experimentations 2016-17 and 2017-18 were found to be non-significant, which is presented in Appendix-III.

4.6 NUMBER OF FLORETS PER PLANT

The data regarding number of florets per plant during the year 2016-17 and 2017-18 as influenced due to sowing dates and fertilizer doses are presented in Table 7 (a), (b) and depicted in Fig. 7.

4.6.1 Effect of sowing dates

The data furnished in Table 7 (a) showed that, the influence of sowing dates on florets per plant was found statistically significant.

In the year 2016-17, the sowing date 10th October (D_1) measured significantly the maximum florets per plant (34.44) and which was found to be at par with the treatment D_2 (33.20). Whereas, the minimum florets per plant (17.45) was counted in the sowing date 20th January (D_6).

During the year 2017-18, significantly the maximum florets per plant (36.47) was observed in the sowing date 10th October (D_1) and which was found to be at par with the treatment D_2 (34.95). However, the sowing date 20th January (D_6) was recorded minimum (16.04) florets per plant.

The pooled data indicated that, number of florets per plant were significantly the maximum (35.45) in treatment D_1. However, the treatment D_6 was recorded significantly the minimum (16.75) florets per plant. The kasuri methi crop sown earlier might had benefited by favorable conditions like temperature, humidity. Ultimately, it achieved better seedling vigour, maximum photosynthetic surface leads to accumulation of maximum fresh

Table 7 (a). Effect of dates of sowing and fertilizer doses on number of florets per plant in kasuri methi

Treatments	Number of florets per plant		
	2016-17	2017-18	Pooled
Dates of sowing (D)			
D_1 - 10th October	34.44	36.47	35.45
D_2 - 30th October	33.20	34.95	34.07
D_3 - 20th November	28.56	25.29	26.93
D_4 - 10th December	24.98	24.58	24.78
D_5 - 30th December	19.40	21.09	20.25
D_6 - 20th January	17.45	16.04	16.75
'F' test	Sig.	Sig.	Sig.
SE(m) ±	0.50	0.55	0.39
CD at 5%	1.43	1.57	1.12
Fertilizer dose (F)			
F_1 - 40:40:20 kg ha^{-1} NPK	24.77	25.02	24.89
F_2 - 40:50:25 kg ha^{-1} NPK	26.93	26.93	26.93
F_3 - 50:40:20 kg ha^{-1} NPK	25.72	25.96	25.84
F_4 - 50:50:25 kg ha^{-1} NPK	27.93	27.70	27.82
'F' test	Sig.	Sig.	Sig.

SE(m) ±	0.41	0.45	0.32
CD at 5%	1.17	1.28	0.91
Interaction (D X F)			
'F' test	Sig.	Sig.	Sig.
SE(m) ±	1.01	1.10	0.79
CD at 5%	2.87	3.13	2.24

Table 7 (b). Interaction effects of dates of sowing and fertilizer doses on number of florets per plant in kasuri methi

Treatments	Number of florets per plant		
	2016-17	**2017-18**	**Pooled**
D_1F_1	32.78	34.95	33.86
D_1F_2	35.47	37.10	36.28
D_1F_3	33.85	36.15	35.00
D_1F_4	35.66	37.68	36.67
D_2F_1	32.04	33.77	32.91
D_2F_2	33.65	35.63	34.64
D_2F_3	33.28	34.60	33.94
D_2F_4	33.81	35.79	34.80
D_3F_1	27.52	25.80	26.66
D_3F_2	28.58	27.72	28.15
D_3F_3	28.50	19.67	24.09
D_3F_4	29.64	27.98	28.81
D_4F_1	23.13	22.34	22.73
D_4F_2	25.44	24.17	24.81
D_4F_3	23.22	26.63	24.93
D_4F_4	28.11	25.17	26.64
D_5F_1	13.95	18.73	16.34
D_5F_2	21.29	20.51	20.90
D_5F_3	20.14	23.05	21.59
D_5F_4	22.24	22.08	22.16
D_6F_1	19.20	14.52	16.86
D_6F_2	17.15	16.48	16.81
D_6F_3	15.31	15.66	15.49

D_6F_4	18.13	17.52	17.83
'F' test	Sig.	Sig.	Sig.
SE(m) ±	1.01	1.10	0.79
CD at 5%	2.87	3.13	2.24

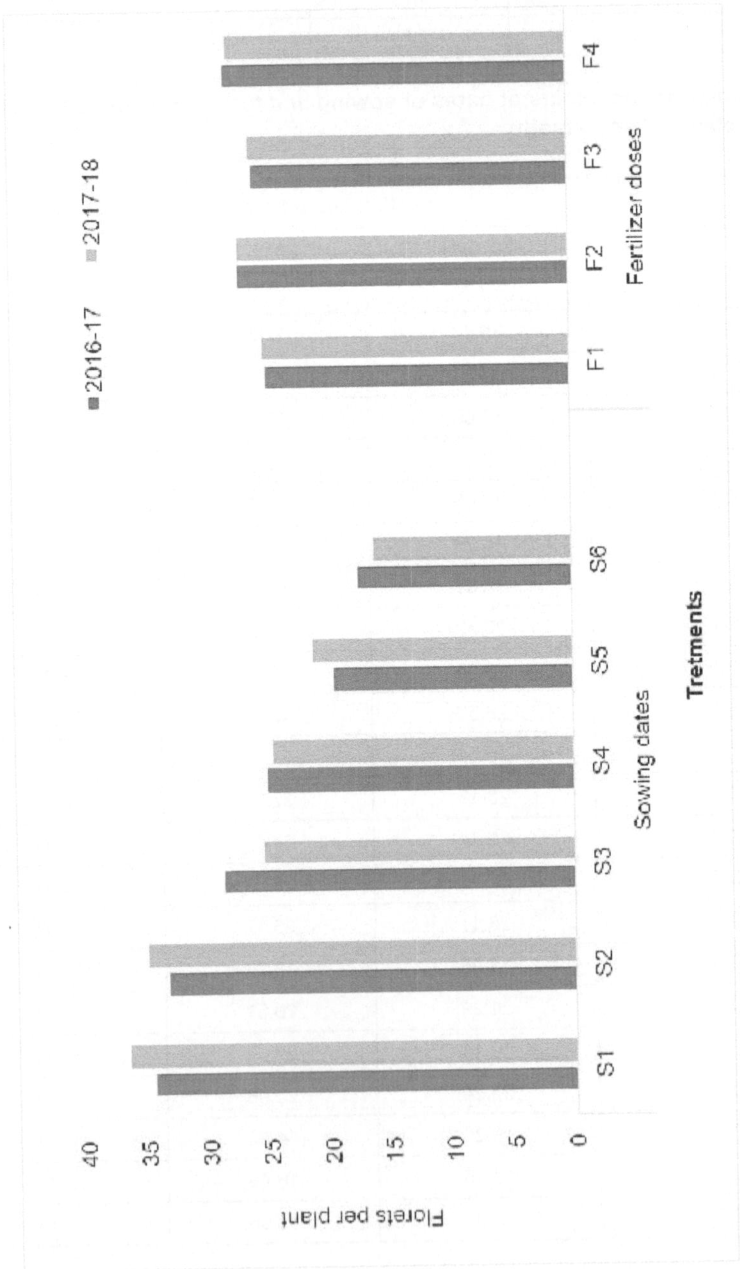

Fig. 7. Effect of dates of sowing and fertilizer doses on number of florets per plant in kasuri methi

Plate 5. Number of florets per plant in kasuri methi crop sown on 10th October along with 50:50:25 kg ha^{-1} NPK (D_1F_4) in the year of 2017-18

Plate 6. Number of florets per plant in kasuri methi crop sown on 20th January with 40:40:20 kg ha^{-1} NPK (D_6F_1) in the year of 2017-18

weight and which might lead into the maximum number of florets as compared to those sown late (20th January). Similar results and the trends were recorded by the earlier workers, Anitha *et al.* (2016) in fenugreek.

4.6.2 Effect of fertilizer doses

The data presented in Table 7 (a) noticed that, the influence of different fertilizer doses in respect of florets per plant were found to be significant during both the years of experimentations.

During the year 2016-17 and 2017-18, the fertilizer dose 50:50:25 kg/ha NPK (F_4) was recorded significantly the maximum number of florets per plant (27.93 and 27.70, respectively) and which was found to be at par with the treatment F_2 (26.93), during both the years of experimentations. However, the fertilizer dose 40:40:20 kg/ha NPK (F_1) showed that, significantly the minimum number of florets per plant (24.77 and 25.02, respectively) were produced in kasuri methi crop.

The pooled data indicated that, the fertilizer dose 50:50:25 kg/ha NPK (F_4) observed significantly the maximum (27.82) florets per plant. Whereas, the fertilizer dose 40:40:20 kg/ha NPK (F_1) was noticed, significantly the minimum (24.89) florets per plant. The kasuri methi crop fertilized with higher level of nutrients reflected into the maximum number of florets per plant as compare to reduced level of NPK in the present study. The reason behind this might be the positive response of more dose of primary nutrients being a major structural element of cell and helped in cell elongation, which in terms develop reproductive structure like florets in kasuri methi. The similar increase in yield attributes were reported by Koyani *et al.* (2012) in fennel.

4.6.3 Interaction effect

The data presented in Table 7 (b) exhibited that, the interaction effect due to sowing dates and fertilizer doses towards the number of florets per plant during both the years of experimentations were found to be statistically significant.

In the year 2016-17, the treatment combination D_1F_4 was recorded significantly the maximum florets per plant (35.66) in kasuri methi. However, the minimum florets per plant (13.95) were recorded in the treatment combination D_5F_1.

Similarly, during the year 2017-18, significantly the maximum florets per plant (37.68) in kasuri methi were observed in the treatment combination D_1F_4. However, the treatment combination D_5F_1 had recorded minimum florets per plant (18.73).

The pooled data indicated that, significantly the maximum florets per plant (36.67) in kasuri methi were recorded in treatment combination D_1F_4. Whereas, the treatment combination D_5F_1 had recorded the minimum florets per plant (16.34).

The kasuri methi crop sown earlier and fertilized with the maximum level of primary nutrients exhibited to maximum number of florets per plant. This might be due to the fact that, earlier sowing crop gets the maximum period for establishment of vegetative growth i.e. plant height, number of branches, number of leaves etc.

Simultaneously, higher level of primary nutrients might also get sufficient time to get available to the root zone of kasuri methi crop. Due to this combine effect; the number of florets per plant might be increased. These findings are in harmony with the findings of Kumar *et al.* (2015) in fenugreek.

4.7 NUMBER OF PODS PER FLORET

The data in respect to the number of pods per floret during the years 2016-17 and 2017-18 as influenced by sowing dates and fertilizer doses are presented in Table 8 (a), (b) and depicted in Fig. 8.

4.7.1 Effect of sowing dates

The data presented in Table 8 (a) revealed that, an effect of different dates of sowing on pods per floret in kasuri methi were found to be statistically significant.

In the year 2016-17, the sowing date 10th October (D_1) was recorded significantly the maximum (23.20) pods per floret in kasuri methi. Whereas, the sowing date 20th January (D_6) was recorded significantly the minimum (11.68) pods per florets.

Table 8 (a). Effect of dates of sowing and fertilizer doses on number of pods per floret in kasuri methi

Treatments	Number of pods per floret		
	2016-17	2017-18	Pooled
Dates of sowing (D)			
D_1 - 10th October	23.20	23.88	23.54
D_2 - 30th October	22.16	23.32	22.74
D_3 - 20th November	19.70	20.48	20.09
D_4 - 10th December	17.39	17.94	17.66
D_5 - 30th December	13.29	15.93	14.61
D_6 - 20th January	11.68	14.27	12.98
'F' test	Sig.	Sig.	Sig.
SE(m) ±	0.32	0.33	0.22
CD at 5%	0.91	0.94	0.62
Fertilizer dose (F)			
F_1 - 40:40:20 kg ha^{-1} NPK	16.92	18.27	17.60
F_2 - 40:50:25 kg ha^{-1} NPK	18.33	19.71	19.02
F_3 - 50:40:20 kg ha^{-1} NPK	17.76	19.16	18.46
F_4 - 50:50:25 kg ha^{-1} NPK	18.60	20.07	19.34
'F' test	Sig.	Sig.	Sig.

SE(m) ±	0.26	0.27	0.18
CD at 5%	0.74	0.78	0.51
Interaction (D X F)			
'F' test	Sig.	Sig.	Sig.
SE(m) ±	0.64	0.66	0.44
CD at 5%	1.82	1.88	1.24

Table 8 (b). Interaction effect of dates of sowing and fertilizer doses on number of pods per floret in kasuri methi

Treatments	Number of pods per floret		
	2016-17	2017-18	Pooled
D_1F_1	22.22	22.97	22.60
D_1F_2	23.58	24.10	23.84
D_1F_3	23.09	24.01	23.55
D_1F_4	23.92	24.45	24.19
D_2F_1	21.44	22.76	22.10
D_2F_2	22.46	23.46	22.96
D_2F_3	21.96	23.12	22.54
D_2F_4	22.79	23.95	23.37
D_3F_1	18.64	19.33	18.98
D_3F_2	19.98	20.94	20.46
D_3F_3	19.37	20.34	19.86
D_3F_4	20.78	21.29	21.04
D_4F_1	16.33	17.53	16.93
D_4F_2	17.68	18.97	18.33
D_4F_3	17.13	16.02	16.58
D_4F_4	18.40	19.23	18.82
D_5F_1	10.01	12.05	11.03
D_5F_2	15.07	16.56	15.81
D_5F_3	14.36	18.19	16.27
D_5F_4	13.72	16.91	15.32
D_6F_1	12.86	15.00	13.93
D_6F_2	11.20	14.24	12.72
D_6F_3	10.65	13.27	11.96
D_6F_4	12.01	14.58	13.30

'F' test	Sig.	Sig.	Sig.
SE(m) ±	0.64	0.66	0.44
CD at 5%	1.82	1.88	1.24

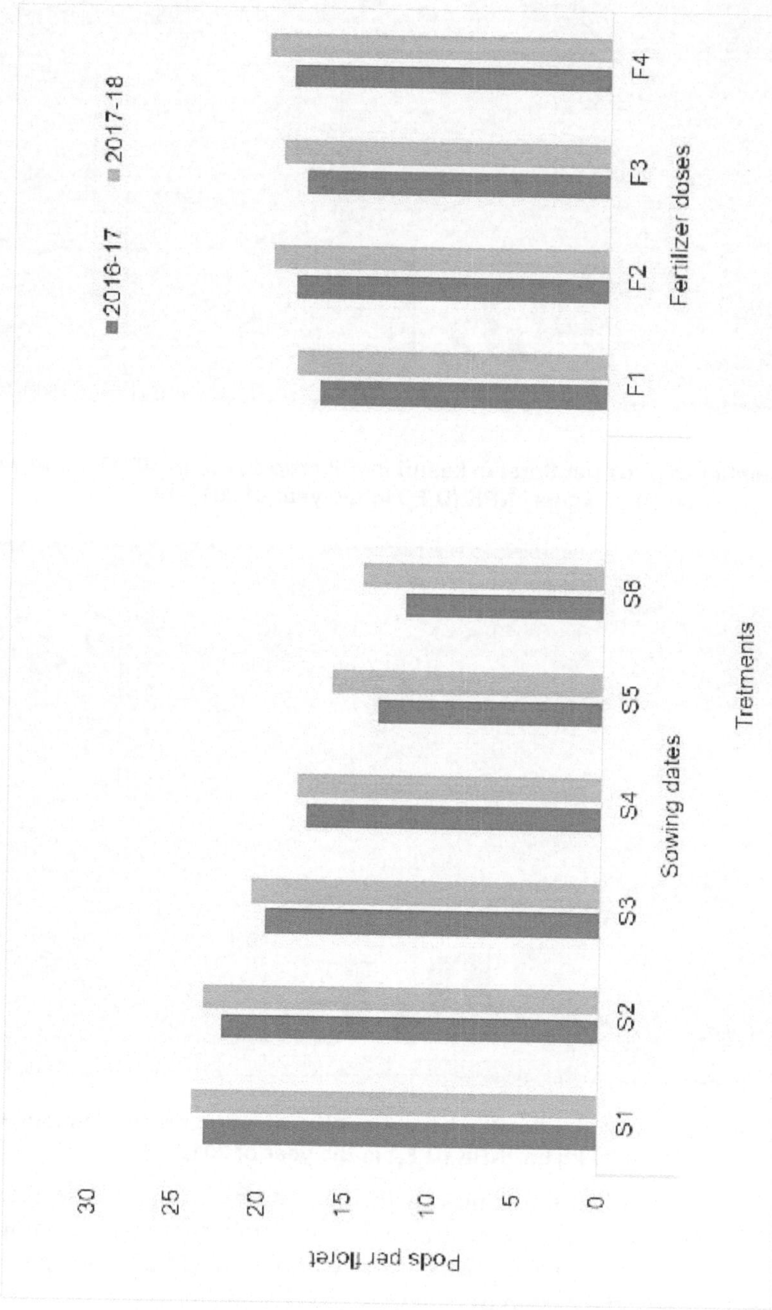

Fig. 8. Effect of dates of sowing and fertilizer doses on number of pods per floret in kasuri methi

Plate 7. Number of pods per floret in kasuri methi crop sown on 10th October with 50:50:25 kg ha^{-1} NPK (D_1F_4) in the year of 2017-18

Plate 8. Number of pods per floret in kasuri methi crop sown on 20th January with 40:40:20 kg ha^{-1} NPK (D_6F_1) in the year of 2017-18

During the year 2017-18, significantly the maximum (23.88) pods per floret in kasuri methi were recorded in the treatment D_1 (10th October) and which was found to be at par with the treatment D_2 (23.32), while minimum (14.27) pods per floret were in kasuri methi observed in treatment D_6 (20th January).

The pooled data furnished in Table 8 (a) found that, significantly the maximum (23.54) pods per floret were observed in kasuri methi crop sown on 10th October (D_1). However, the treatment D_6 was recorded significantly the minimum (12.98) pods per floret. The early sown kasuri methi crop produced the maximum pods per floret might be due to better start of vegetative growth, side by side the favorable and congenial micro-climate inside the soil. The early sown kasuri methi crop produced more number of branches, leaves etc. Thereby naturally accumulation of more photosynthets, which in terms might be resulted in better number of pods per florets. These results are in line with the findings of Meena *et al.* (2018) and Kauser *et al.* (2018) in fenugreek.

4.7.2 Effect of fertilizer doses

The data presented in Table 8 (a) noticed that, the influence of different doses of fertilizer in respect of number of pods per florets were found to be significant, during both the years of experimentations.

During the year 2016-17 and 2017-18, the fertilizer dose 50:50:25 kg/ha NPK (F_4) was recorded significantly the maximum pods per floret (18.60 and 20.07, respectively) and which was found to be at par with the treatment F_2 (18.33 and 19.71, respectively). However, fertilizer dose 40:40:20 kg/ha NPK (F_1) was showed significantly the minimum (16.92 and 18.27, respectively) pods per floret in kasuri methi crop.

The pooled data indicated that, the fertilizer dose 50:50:25 kg/ha NPK (F_4) was produced significantly the maximum (19.34) pods per floret. Whereas, the fertilizer dose 40:40:20 kg/ha NPK (F_1) noticed significantly the minimum (17.60) pods per floret. The higher level of primary nutrients were found to be effective in more vegetative growth as well as reproductive growth in terms of florets per plant. This might promote better photosynthetic activities, resulted in increased carbohydrates synthesis, which might resulted in increased number of pods per floret. This findings are in conformity with the results of Sunanda *et al.* (2014) in kasuri methi and Mitoo *et al.* (2018) in fenugreek.

4.7.3 Interaction effect

The data presented in Table 8 (b) exhibited that, an interaction effect due to sowing dates and fertilizer doses on the number of pods per floret during the years 2016-17 and 2017-18 were found to be statistically significant.

The treatment combination D_1F_4 was recorded significantly the maximum pods per floret in 2016-17 (23.92) and 2017-18 (24.45). Whereas, the treatment combination D_6F_3 was recorded minimum pods per floret in 2016-17 (10.65) and 2017-18 (13.27) in kasuri methi crop.

The pooled data indicated that, significantly the maximum pods per floret (24.19) in kasuri methi was recorded in treatment combination D_1F_4. Whereas, the treatment combination D_6F_3 had recorded the minimum pods per floret (11.96).

More number of pods per floret is a supportive characters for more seed yield production in kasuri methi crop. Early sowing embedded with higher level of primary nutrients might be responsible for getting the more period of time to the kasuri methi crop regarding temperature, relative humidity and soil micro-temperature, which would be more helpful to the pod setting. The favourable agro-climatic conditions exerted in the production of more number of pods per floret in kasuri methi. The results of the present investigation are supported by the findings of Bhardwaj (1996) in french bean and Bhutiya and Sharangi (2016) in fenugreek.

4.8 NUMBER OF SEEDS PER POD

The data in respect of number of seeds per pod in kasuri methi during the years 2016-17 and 2017-18 as influenced by sowing dates and fertilizer doses were presented in Table 9.

Table 9. Effect of dates of sowing and fertilizer doses on number of seeds per pod in kasuri methi

Treatments	Number of seeds per pod		
	2016-17	2017-18	Pooled
Dates of sowing (D)			
D_1 - 10th October	5.79	5.47	5.63
D_2 - 30th October	5.59	5.23	5.41
D_3 - 20th November	4.54	4.81	4.67
D_4 - 10th December	3.81	4.27	4.04
D_5 - 30th December	2.91	3.38	3.15
D_6 - 20th January	2.49	3.35	2.92
'F' test	Sig.	Sig.	Sig.
SE(m) ±	0.07	0.09	0.06
CD at 5%	0.19	0.25	0.17
Fertilizer dose (F)			
F_1 - 40:40:20 kg ha^{-1} NPK	3.85	4.21	4.03
F_2 - 40:50:25 kg ha^{-1} NPK	4.28	4.43	4.36
F_3 - 50:40:20 kg ha^{-1} NPK	4.19	4.42	4.31
F_4 - 50:50:25 kg ha^{-1} NPK	4.43	4.62	4.52
'F' test	Sig.	Sig.	Sig.
SE(m) ±	0.05	0.07	0.05

CD at 5%	0.16	0.21	0.14
Interaction (D X F)			
'F' test	Sig.	NS	NS
SE(m) ±	0.13	0.18	0.12
CD at 5%	0.38	-	-

4.8.1 Effect of sowing dates

The data presented in Table 9 revealed that, the different dates of sowing were found to be significant during both the years of experiments as well as in calculated pooled results.

In the year 2016-17, the sowing date 10th October (D_1) recorded significantly the maximum (5.79) number of seeds per pod. However, significantly the minimum (2.49) seeds per pod were found in sowing date 20th January (D_6).

During the year 2017-18, significantly the maximum (5.47) number of seeds per pod were produced in the sowing date 10th October (D_1) and which was found to be at par with the treatment D_2 (5.23). Whereas, the sowing date 20th January (D_6) was recorded significantly the minimum (3.35) number of seeds per pod in kasuri methi crop.

The pooled data furnished in Table 9 said that, the sowing date 10th October (D_1) was recorded significantly the maximum (5.63) number of seeds per pod. Whereas, sowing date 20th January (D_6) was recorded minimum (2.92) seeds per pod. With the advancement of sowing date, there where decreasing trend of seeds per pod in kasuri methi was noticed in the present investigation. The earlier sowing of kasuri methi crop produces the maximum number of seeds per pod. This might be owing to the fact that, the early sown kasuri methi crop had the benefit of full season of growth and better atmospheric conditions, whereas, later sown kasuri methi crop had a shorter growth phase. Furthermore, the later sown kasuri methi crop get higher temperature and less humidity, might be resulted in less number of seeds per pod. The results of the present investigation were supported by the findings of Bhardwaj (1996) in french bean and Anupama *et al.* (2017) in kasuri methi.

4.8.2 Effect of fertilizer doses

The data furnished in Table 9 revealed that, the fertilizer dose 50:50:25 kg/ha NPK (F_4) was recorded significantly the maximum (4.43) number of seeds per pod and which was found to be at par with the treatment F_2 (4.28). Whereas, fertilizer dose 40:40:20 kg/ha NPK (F_1) was recorded minimum (3.85) number of seeds per pod in kasuri methi crop during the year 2016-17.

In the year 2017-18, the fertilizer dose 50:50:25 kg/ha NPK (F_4) was recorded significantly the maximum (4.62) number of seeds per pod and which was found to be at par with the treatment F_2 (4.43) and F_3 (4.42). However, the fertilizer dose

40:40:20 kg/ha NPK (F_1) was recorded significantly the minimum (4.21) number of seeds per pod in kasuri methi crop.

The pooled data presented in Table 9 revealed that, the treatment F_4 was recorded significantly the maximum (4.52) number of seeds per pod. However, significantly minimum (4.03) seeds per pod were observed in treatment F_1. This might be due to fact that, the supply of increased levels of primary nutrients i.e. nitrogen increased concentration of carbohydrates and the seed could be served reservoir for the carbohydrates, the phosphorus govern the metabolic activities of plant thereby, increases number of seeds per pod and finally potassium took responsibility of sequential metamorphosis from the chain of source to sink relationship. Ultimately, it resulted into the maximum number of seeds per pod in kasuri methi as compared to rest of fertilizer levels. Similar results have been recorded by the earlier workers like Bhardwaj (1996) in french bean and Sunanda *et al.* (2014) in kasuri methi.

4.8.3 Interaction effect

The data presented in Appendix-IV revealed that, the interaction effects between sowing dates and fertilizer levels were found to be statistically significant during 2016-17. However, it was noticed statistically non-significant results in the present investigation during 2017-18 and pooled data analysis too in kasuri methi crop.

4.9 SEED YIELD PER PLANT (G)

The data in respect of seed yield per plant during the years 2016-17 and 2017-18 as influenced by sowing dates and fertilizer doses are presented in Table 10 (a), (b) and depicted in Fig. 9.

4.9.1 Effect of sowing doses

The data presented in Table 10 (a) revealed that, an effect of different sowing dates on seed yield per plant were found to be statistically significant.

In the year 2016-17, the treatment D_1 (10th October) was recorded significantly the maximum (5.30 g) seed yield per plant and which was found to be at par with the treatment D_2 (5.17). Whereas, the treatment D_6 (20th January) was recorded significantly the minimum (2.71 g) seed yield per plant.

During the year 2017-18, significantly the maximum (5.34 g) seed yield per plant was recorded in the treatment D_1 and which was found to be at par with the treatment D_2 (5.19). While, significantly the minimum (2.68 g) seed yield per plant was observed in treatment D_6 (20th January) in kasuri methi crop.

The pooled data furnished in Table 10 (a) found that, significantly the maximum (5.32 g) seed yield per plant was observed in treatment D_1 (10th October) and which was found to be at par with D_2 (5.18). However, the treatment D_6 (20 January) was recorded significantly the minimum (2.69 g) seed yield per plant in kasuri methi crop. This might be due to the fact that, the higher seed yield per plant with an early sowing (10th October) could be attributed due to its beneficial influence on yield attributes, because the crop had longer growth period and availability of favourable soil moisture and temperature during crop growth period. The decline in seed yield per plant in later sown crop might be due to the shorter duration of reproductive phase coinciding with increasing temperature which could have allowed less period of time for development of seed. Similar results have been recorded by the earlier workers, Nandre *et al.* (2011) in fenugreek and Sharma (2006) in kasuri methi.

4.9.2 Effect of fertilizer doses

The data presented in Table 10 (a) noticed that, an influence of different fertilizer doses in respect of seed yield per plant was found to be significant during both the years of experimentations.

Table 10 (a). Effect of dates of sowing and fertilizer doses on seed yield per plant (g) in kasuri methi

Treatments	Seed yield per plant (g)		
	2016-17	2017-18	Pooled
Dates of sowing (D)			
D_1 - 10th October	5.30	5.34	5.32
D_2 - 30th October	5.17	5.19	5.18
D_3 - 20th November	4.36	4.32	4.34
D_4 - 10th December	3.71	3.51	3.61
D_5 - 30th December	3.16	3.36	3.26
D_6 - 20th January	2.71	2.68	2.69
'F' test	Sig.	Sig.	Sig.
SE(m) ±	0.05	0.06	0.05
CD at 5%	0.14	0.18	0.15
Fertilizer dose (F)			
F_1 - 40:40:20 kg ha^{-1} NPK	3.91	3.89	3.90
F_2 - 40:50:25 kg ha^{-1} NPK	4.12	4.12	4.12
F_3 - 50:40:20 kg ha^{-1} NPK	3.98	3.99	3.99
F_4 - 50:50:25 kg ha^{-1} NPK	4.25	4.26	4.26
'F' test	Sig.	Sig.	Sig.
SE(m) ±	0.04	0.06	0.04
CD at 5%	0.12	0.18	0.12
Interaction (D X F)			

'F' test	Sig.	Sig.	Sig.
SE(m) ±	0.10	0.13	0.11
CD at 5%	0.29	0.36	0.30

Table 10 (b). Interaction effect of dates of sowing and fertilizer doses on seed yield per plant (g) in kasuri methi

Treatments	Seed yield per plant (g)		
	2016-17	2017-18	Pooled
D_1F_1	5.16	5.18	5.17
D_1F_2	5.36	5.43	5.39
D_1F_3	5.26	5.29	5.28
D_1F_4	5.40	5.45	5.42
D_2F_1	5.08	5.10	5.09
D_2F_2	5.22	5.24	5.23
D_2F_3	5.13	5.15	5.14
D_2F_4	5.25	5.28	5.26
D_3F_1	4.13	4.12	4.12
D_3F_2	4.44	4.38	4.41
D_3F_3	4.29	4.27	4.28
D_3F_4	4.58	4.52	4.55
D_4F_1	3.79	3.57	3.68
D_4F_2	3.67	3.42	3.55
D_4F_3	3.37	3.09	3.23
D_4F_4	3.99	3.96	3.97
D_5F_1	2.61	3.03	2.82
D_5F_2	3.27	3.36	3.31
D_5F_3	3.36	3.66	3.51
D_5F_4	3.38	3.40	3.39
D_6F_1	2.67	2.36	2.52
D_6F_2	2.76	2.89	2.83
D_6F_3	2.47	2.48	2.48
D_6F_4	2.92	2.97	2.95
'F' test	Sig.	Sig.	Sig.
SE(m) ±	0.10	0.13	0.11
CD at 5%	0.29	0.36	0.30

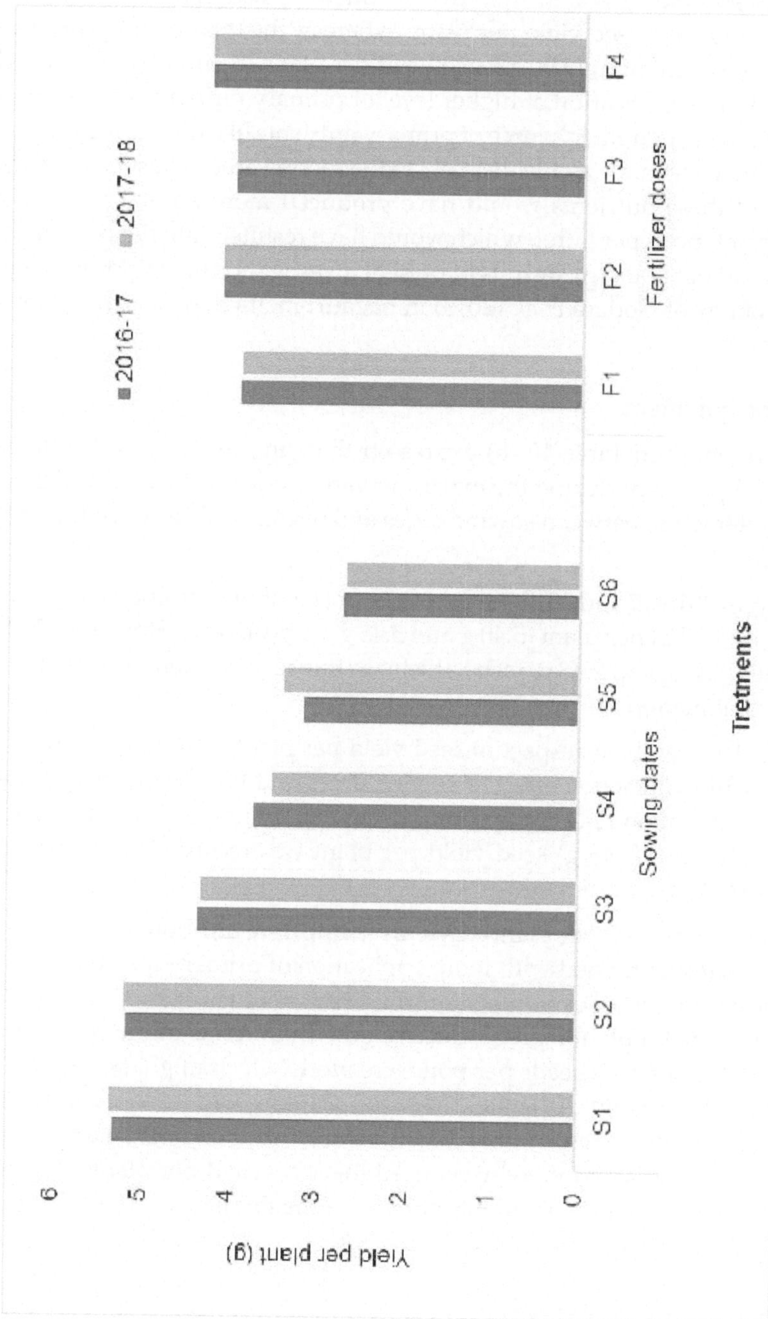

Fig. 9. Effect of dates of sowing and fertilizer doses on seed yield per plant (g) in kasuri methi

During the year 2016-17 and 2017-18, the treatment F_4 (50:50:25 kg/ha NPK) was recorded significantly the maximum seed yield per plant (4.25 and 4.26 g, respectively). However, treatment F_1 (40:40:20 kg/ha NPK) was showed significantly the minimum (3.91 and 3.89 g, respectively) seed yield per plant in kasuri methi crop.

The pooled data indicated that, the treatment F_4 was observed significantly the maximum (4.26 g) seed yield per plant. Whereas, the treatment F_1 was noticed significantly the minimum (3.90 g) seed yield per plant. The maximum seed yield per plant with the application of higher level of primary nutrients in kasuri methi might be due to the inorganic source of primary nutrients like nitrogen, phosphorus and potassium were made available freely to the rooting area of kasuri methi. More availability of these nutrients would have produced more number of branches, florets per plant, pods per floret, which would have resulted into the production of more seed yield per plant (g). Results of present investigation are in close agreement with the findings of Godara *et al.* (2013) in nagauri methi and Bhardwaj (1996) in french bean.

4.9.3 Interaction effect

The data presented in Table 10 (b) expressed that, an interaction effect towards the seed yield per plant during the experimentation years of 2016-17 and 2017-18, due to the interaction between sowing dates and fertilizer doses were found to be significant.

In the year 2016-17 and 2017-18, the treatment combination D_1F_4 was produced maximum seed yield per plant (5.40 g and 5.45 g, respectively). However, the minimum values (2.47 g and 2.48 g, respectively) in this regards were recorded in D_6F_3 treatment combination.

The pooled results in respect of seed yield per plant were significantly influenced due to the interaction effect of sowing dates and fertilizer doses. However, treatment combination D_1F_4 was produced the maximum (5.42 g) seed yield per plant and minimum (2.48 g) seed yield per plant was recorded in the treatment combination D_6F_3 in kasuri methi crop.

The maximum seed per plant in kasuri methi were attributed due to the fact that, early sowing combined with more application of primary nutrients, resulted in vigorous growth and development of the plant. Because of such strong framework of kasuri methi plant its reproductive growth in terms of number of florets per plant, pods per floret, seeds per pod were more, which might be resulted into more seed yield per plant. As against, delayed planting with reduced dose of primary nutrients might be responsible for shortening the phonological development of kasuri methi plants and thereby reduced the seed yield per plant. The results of the present investigation are in harmony with the findings of Ali *et al.* (2015) in coriander and Sheoran *et al.* (2000) in fenugreek.

4.10 SEED YIELD PER PLOT (G)

The data in respect of seed yield per plot during the year 2016-17 and 2017-18, as influenced by sowing dates and fertilizer doses are presented in Table 11, 12 (b) and depicted in Fig. 10.

4.10.1 Effect of sowing dates

The data presented in Table 11 revealed that, the different sowing dates and fertilizer doses were found to be statistically significant during both the years of experiments as well as in pooled results too.

In the year 2016-17, the sowing date 10th October (D_1) was recorded significantly the maximum (367.99 g) seed yield per plot and which was found to be at par with the treatment D_2 (360.29 g). However, significantly the minimum (191.24 g) seed yield per plot was found in sowing date 20th January (D_6) in kasuri methi crop.

During the year 2017-18, significantly the maximum (373.55 g) seed yield per plot was observed in sowing date 10th October (D_1) and which was found to be at par with the treatment D_2 (363.41 g). Whereas, the sowing date 20th January (D_6) was recorded significantly the minimum (187.54 g) seed yield per plot in kasuri methi crop.

The pooled data furnished in Table 11 observed that, the sowing date 10th October (D_1) was recorded significantly the maximum (370.77 g).

Table 11. Effect of dates of sowing and fertilizer doses on seed yield per plot (g) in kasuri methi

Treatments	Seed yield per plot (g)		
	2016-17	2017-18	Pooled
Dates of sowing (D)			
D_1 - 10th October	367.99	373.55	370.77
D_2 - 30th October	360.29	363.41	361.85
D_3 - 20th November	307.99	302.31	305.15
D_4 - 10th December	273.25	245.60	259.42
D_5 - 30th December	206.26	235.51	220.88
D_6 - 20th January	191.24	187.54	189.39
'F' test	Sig.	Sig.	Sig.
SE(m) ±	4.99	4.41	3.54
CD at 5%	14.17	12.56	10.08
Fertilizer dose (F)			
F_1 - 40:40:20 kg ha^{-1} NPK	274.61	272.54	273.58
F_2 - 40:50:25 kg ha^{-1} NPK	288.25	288.41	288.33
F_3 - 50:40:20 kg ha^{-1} NPK	278.07	279.30	278.69
F_4 - 50:50:25 kg ha^{-1} NPK	297.08	298.35	297.71
'F' test	Sig.	Sig.	Sig.
SE(m) ±	4.07	3.60	2.89
CD at 5%	11.57	10.26	8.23

Interaction (D X F)			
'F' test	Sig.	Sig.	Sig.
SE(m) ±	9.96	8.83	7.08
CD at 5%	28.35	25.12	20.16

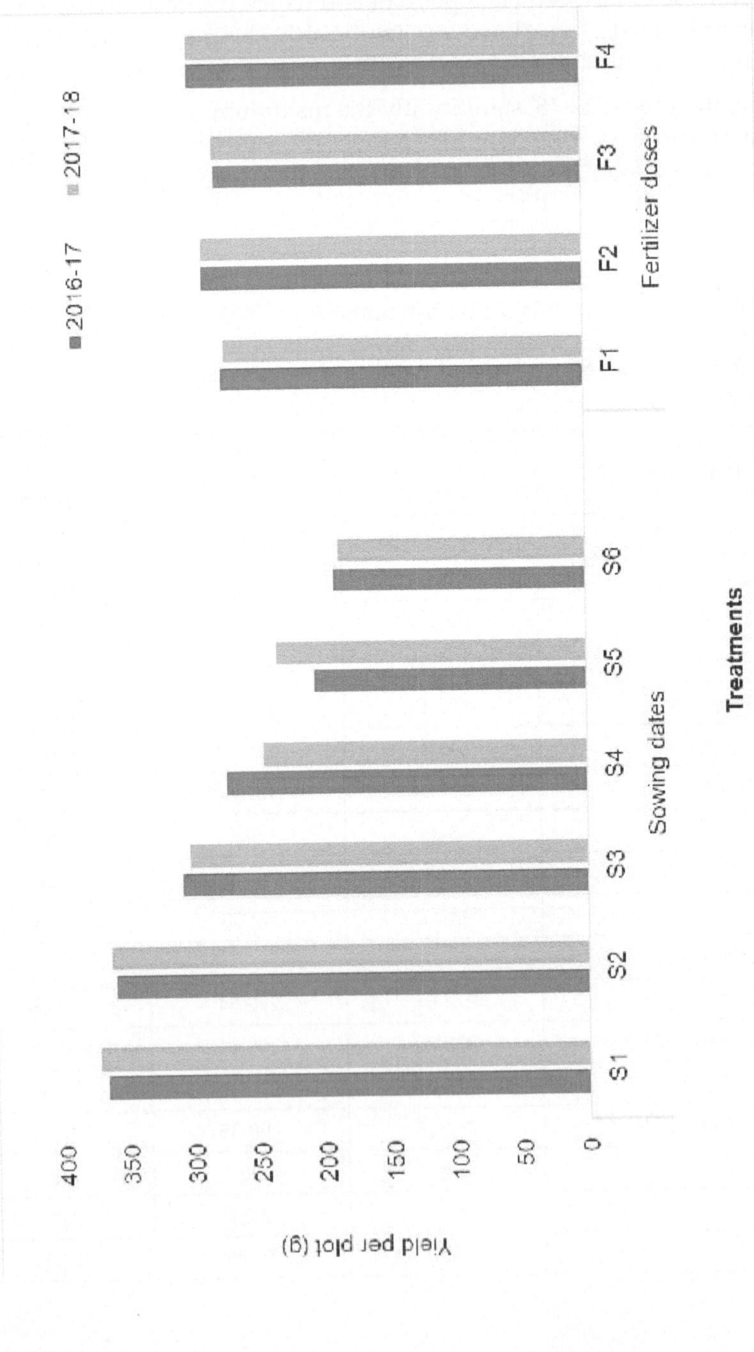

Fig. 10. Effect of dates of sowing and fertilizer doses on seed yield per plot (g) in kasuri methi

seed yield per plot found to be at par with D_2 (361.85 g). Whereas, sowing date 20th January (D_6) was recorded minimum (189.39 g) seed yield per plot in kasuri methi crop.

This might be due to the improvement in overall growth i.e. better vegetative growth with an early sowing date (10th October) coupled with increased net photosynthesis towards reproduction structure, on the other hand, such crop might have increase the yield attributes like florets per plant, pods per floret and seeds per pod, which might be turned into more seed yield per plot. Similar results have been recorded by the earlier workers like Nandre *et al.* (2011) and Singh *et al.* (2017[a]) in fenugreek.

4.10.2 Effect of fertilizer doses

The data furnished in Table 11 revealed that, the fertilizer dose 50:50:25 kg/ha NPK (F_4) was recorded significantly the maximum (297.08 g) seed yield per plot and which was found to be at par with the treatment F_2 (288.25 g). Whereas, the fertilizer dose 40:40:20 kg/ha NPK (F_1) was recorded minimum (274.61 g) seed yield per plot, during the year 2016-17.

In the year 2017-18, the fertilizer dose 50:50:25 kg/ha NPK (F_4) was recorded significantly the maximum (298.35 g) seed yield per plot and which was found to be at par with the treatment F_2 (288.41 g). However, the fertilizer dose 40:40:20 kg/ha NPK (F_1) was recorded significantly the minimum (272.54 g) seed yield per plot.

The pooled data presented in Table 11 noticed that, the fertilizer dose 50:50:25 kg/ha NPK (F_4) was recorded significantly the maximum (297.71 g) seed yield per plot. However, significantly the minimum (273.58 g) seed yield was observed in the fertilizer dose 40:40:20 kg/ha NPK (F_1) in kasuri methi crop.

The application of higher level of primary nutrients might have positive effects on growth parameters like plant height and number of branching which might have passed through long span and this might be resulted to delay in maturity and ultimately resulted in yields and yield attributes in kasuri methi seeds. Which is positively correlated with number of branches per plant, florets per plant, pods per floret and seeds per pod in kasuri methi crop. Similar results have been recorded by the earlier workers like Datta *et al.* (2017) in fenugreek.

4.10.3 Interaction effect

The data presented in Table 12 (b) had showed that, an interaction effect regarding the seed yield per plot during both the years of experimentations due to the interaction between sowing dates and fertilizer doses were found to be significant in kasuri methi crop.

In the year 2016-17, the treatment combination D_1F_4 was recorded significantly the maximum (374.97 g) seed yield per plot. However, significantly the minimum (153.30 g) seed yield per plot was found in the treatment combination D_5F_1.

During the year 2017-18, significantly the maximum (381.31 g) seed yield per plot was observed in treatment combination D_1F_4. Whereas, the treatment combination D_6F_1 was recorded significantly the minimum (165.52 g) seed yield per plot.

The pooled result in respect of seed yield per plot was significantly influenced due to an interaction effect of dates of sowing and fertilizer doses. The treatment combination D_1F_4 was produced maximum (378.14 g) seed yield per plot and minimum (172.66 g) seed yield per plot was recorded in the treatment combination D_6F_3.This could be attributed due to the fact that, in early sown kasuri methi crop might had benefited by favourable conditions like temperature and humidity could achieved better seedling vigour, maximum photosynthetic surface which might leads to accumulation to the maximum fresh and dry weight as compare to late sown kasuri methi crop. Similarly, with an application of higher level of nitrogen, phosphorus and potash increased root mass. Due to application of nitrogen is largely responsible for increased plant uptake of phosphorus. At the same time higher application of phosphorus might responsible for early root growth and proliferation, which might resulted into desired nitrogen uptake. Hence, application of nitrogen and phosphorus in conjugation might resulted into maximum florets per plant, pods per florets and seeds per pod and ultimately, the seed yield per plot. Similar results have been recorded by the earlier workers like Anitha *et al.* (2016) and Meena *et al.* (2018) in fenugreek.

4.11 SEED YIELD PER HECTARE (Q)

The data in respect of seed yield per hectare during the years 2016-17 and 2017-18 as influenced by sowing dates and fertilizer doses are presented in Table 12 (a), (b) and depicted in Fig 11 in kasuri methi crop.

4.11.1 Effect of sowing dates

The data presented in Table 12 (a) revealed that, an effect of different sowing dates on seed yield per hectare were found to be statistically significant.

In the year 2016-17, the sowing date 10th October (D_1) was recorded significantly the maximum (8.76 q) seed yield per hectare and which was found to be at par with the treatment D_2 (8.58 q). Whereas, the sowing date 20th January (D_6) was recorded significantly the minimum (4.55 q) seed yield per hectare.

During the year 2017-18, significantly the maximum (8.89 q) seed yield per hectare was recorded in the sowing date 10th October (D_1) and which was found to be at par with D_2 (8.65 q). While, significantly the minimum (4.47 q) seed yield per hectare was observed in treatment of sowing date 20th January (D_6).

The pooled data furnished in Table 12 (a) found that, significantly the maximum (8.83 q) seed yield per hectare was obtained in sowing date 10th October (D_1). However, sowing date 20th January (D_6) was recorded significantly the minimum (4.51 q) seed yield per hectare. It is well known fact that, more the seed yield per plot of kasuri methi seed, when calculating simply with the help of hectare factor one could get the maximum seed yield per hectare in the present investigation. Similar results have been recorded by the earlier workers, Anitha *et al.* (2016) and Mitoo *et al.* (2018) in fenugreek.

4.11.2 Effect of fertilizer doses

The data presented in Table 12 (a) noticed that, an influence of different fertilizer doses on seed yield per hectare was found to be

Table 12 (a). Effect of dates of sowing and fertilizer doses on seed yield per hectare (q) in kasuri methi

Treatments	Seed yield per ha (q)		
	2016-17	2017-18	Pooled
Dates of sowing (D)			
D_1 - 10th October	8.76	8.89	8.83
D_2 - 30th October	8.58	8.65	8.42
D_3 - 20th November	7.33	7.20	7.27
D_4 - 10th December	6.51	5.85	6.18
D_5 - 30th December	4.91	5.61	5.27
D_6 - 20th January	4.55	4.47	4.51
'F' test	Sig.	Sig.	Sig.
SE(m) ±	0.12	0.11	0.08
CD at 5%	0.34	0.30	0.24
Fertilizer dose (F)			
F_1 - 40:40:20 kg ha^{-1} NPK	6.54	6.49	6.51
F_2 - 40:50:25 kg ha^{-1} NPK	6.86	6.87	6.87
F_3 - 50:40:20 kg ha^{-1} NPK	6.62	6.65	6.64
F_4 - 50:50:25 kg ha^{-1} NPK	7.07	7.10	7.09
'F' test	Sig.	Sig.	Sig.
SE(m) ±	0.10	0.09	0.07
CD at 5%	0.28	0.24	0.20
Interaction (D X F)			
'F' test	Sig.	Sig.	Sig.
SE(m) ±	0.24	0.21	0.17
CD at 5%	0.68	0.60	0.48

Table 12 (b). Interaction effect of dates of sowing and fertilizer doses on seed yield per plot (g) and per hectare (q) in kasuri methi

Treatments	Seed yield per plot (g)			Seed yield per hectare (q)		
	2016-17	2017-18	Pooled	2016-17	2017-18	Pooled
D_1F_1	360.31	362.67	361.49	8.58	8.64	8.61
D_1F_2	370.56	379.94	375.25	8.82	9.05	8.93
D_1F_3	366.14	370.29	368.22	8.72	8.82	8.77
D_1F_4	374.97	381.31	378.14	8.93	9.08	9.00
D_2F_1	354.43	356.96	355.70	8.44	8.50	8.47
D_2F_2	363.14	367.06	365.10	8.65	8.74	8.69
D_2F_3	358.57	360.19	359.38	8.54	8.58	8.56
D_2F_4	365.00	369.40	367.20	8.69	8.80	8.74
D_3F_1	289.48	288.21	288.85	6.89	6.86	6.88
D_3F_2	315.17	306.32	310.74	7.50	7.29	7.40
D_3F_3	301.71	298.62	300.16	7.18	7.11	7.15
D_3F_4	325.62	316.07	320.85	7.75	7.53	7.64
D_4F_1	281.18	249.70	265.44	6.69	5.95	6.32
D_4F_2	274.20	239.51	256.85	6.53	5.70	6.12
D_4F_3	256.02	216.29	236.15	6.10	5.15	5.62
D_4F_4	281.61	276.88	279.25	6.71	6.59	6.65
D_5F_1	153.30	212.20	182.75	3.65	5.05	4.35
D_5F_2	222.54	235.06	228.80	5.30	5.60	5.45
D_5F_3	214.56	256.53	235.55	5.11	6.11	5.61
D_5F_4	234.63	238.25	236.44	5.59	5.67	5.63
D_6F_1	208.97	165.52	187.24	4.98	3.94	4.46
D_6F_2	183.92	202.55	193.24	4.38	4.82	4.60
D_6F_3	171.43	173.88	172.66	4.08	4.14	4.11

D_6F_4	200.63	208.20	204.42	4.78	4.96	4.87
'F' test	Sig.	Sig.	Sig.	Sig.	Sig.	Sig.
SE(m) ±	9.96	8.83	7.08	0.24	0.21	0.17
CD at 5%	28.35	25.12	20.16	0.68	0.60	0.48

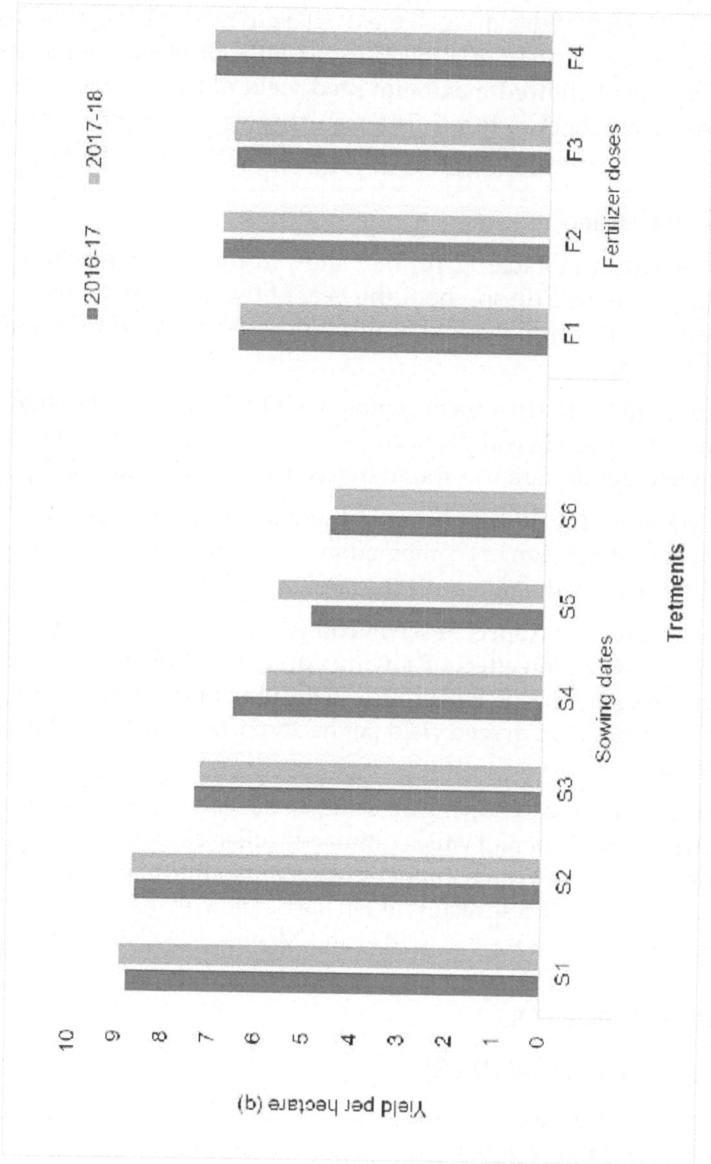

Fig. 11. Effect of dates of sowing and fertilizer doses on seed yield per hectare (q) in kasuri methi

Significant during both the years of experimentations in kasuri methi crop.

During the years 2016-17 and 2017-18, the fertilizer dose 50:50:25 kg/ha NPK (F_4) was recorded significantly the maximum seed yield per hectare (7.07 and 7.10 q, respectively) and which was found to be at par with the treatment F_2 (6.86 and 6.87, respectively). However, fertilizer dose 40:40:20 kg/ha NPK (F_1) was showed significantly the minimum (6.54 and 6.49 q, respectively) seed yield per hectare.

The pooled data indicated that, the fertilizer dose 50:50:25 kg/ha NPK (F_4) was recorded significantly the maximum (7.09 q) seed yield per hectare. Whereas, the fertilizer dose 40:40:20 kg/ha NPK (F_1) was noticed significantly the minimum (6.51 q) seed yield per hectare. When the seed yield of kasuri methi per plot was recorded maximum with an application of 50:50:25 kg/ha NPK. It would be natural thing that, similar treatment showed maximum seed yield of kasuri methi per hectare by multiplying it with hectare factor. The results of the investigation are in close agreement with the findings of Sharma (2006) and Sunanda *et al*. (2014).

4.11.3 Interaction effect

The data presented in in Table 12 (b) had showed that, an interaction effect of the seed yield per hectare during both the years of experimentations due to the interaction between sowing dates and fertilizer doses were found to be significant, in kasuri methi crop.

In the year 2016-17, the treatment combination D_1F_4 was recorded significantly the maximum (8.93 q) seed yield per hectare. However, significantly the minimum (3.65 q) seed yield per hectare was found in treatment combination D_5F_1.

During the year 2017-18, significantly the maximum (9.08 q) seed yield per hectare was observed in treatment combination D_1F_4. Whereas, the treatment combination of D_6F_1 was recorded significantly the minimum (3.94 q) seed yield per hectare.

The pooled results in respect of seed yield per hectare was significantly influenced due to the interaction effect of sowing dates and fertilizer doses. However, the treatment combination D_1F_4 was produced the maximum (9.00 q) seed yield per hectare and minimum (4.11 q) seed yield per hectare was recorded in the treatment combination D_6F_3.

The combination of early sowing date and fertilizer doses at higher levels gave the maximum seed yield per plot which ultimately reflected in its arithmetical calculation in per hectare basis. Which might be observed in the treatment combination D_1F_4 in the present study. These results are in harmony with the findings of Sharma (2006) in kasuri methi, Nandre *et al*. (2011) and Meena *et al*. (2018) in fenugreek.

C. QUALITY PARAMETERS

4.12 TEST WEIGHT OF SEED (G)

The data in respect of test weight during the year 2016-17 and 2017-18 as influenced by sowing dates and fertilizer levels are presented in Table 13.

4.12.1 Effect of sowing doses

The data presented in Table 13 revealed that, the different sowing dates on test weight of kasuri methi seed were found to be significant during both the years of experiments as well as in calculated pooled.

In the year 2016-17, sowing date 10th October (D_1) was recorded significantly the maximum (1.175 g) test weight and which was found to be at par with the treatment D_2 (1.174 g). However, significantly the minimum (1.165 g) test weight was found in treatment of sowing date 20th January (D_6).

During the year 2017-18, significantly the maximum (1.178 g) test weight was observed in sowing date 10th October (D_1) and which was found to be at par with the treatment D_2 (1.177 g). Whereas, the sowing date 20th January (D_6) was recorded significantly the minimum (1.166 g) test weight.

The pooled data furnished in Table 13 said that, the sowing date 10th October (D_1) was recorded significantly the maximum (1.176 g) test weight. Whereas, the crop sown on 20th January (D_1) was recorded minimum (1.166 g) test weight.

Table 13. Effect of dates of sowing and fertilizer doses on test weight of seed (g) in kasuri methi

Treatments	Test weight of seed (g)		
	2016-17	2017-18	Pooled
Dates of sowing (D)			
D_1 - 10th October	1.175	1.178	1.176
D_2 - 30th October	1.174	1.177	1.175
D_3 - 20th November	1.172	1.174	1.13
D_4 - 10th December	1.170	1.171	1.170
D_5 - 30th December	1.166	1.169	1.167
D_6 - 20th January	1.165	1.166	1.166
'F' test	Sig.	Sig.	Sig.
SE(m) ±	0.0005	0.0006	0.0004
CD at 5%	0.0014	0.0016	0.0011
Fertilizer dose (F)			
F_1 - 40:40:20 kg ha^{-1} NPK	1.168	1.171	1.170
F_2 - 40:50:25 kg ha^{-1} NPK	1.171	1.173	1.172
F_3 - 50:40:20 kg ha^{-1} NPK	1.171	1.172	1.171
F_4 - 50:50:25 kg ha^{-1} NPK	1.172	1.173	1.172
'F' test	Sig.	Sig.	Sig.
SE(m) ±	0.0004	0.0005	0.0003

CD at 5%	0.0011	0.0013	0.0009
Interaction (D X F)			
'F' test	NS	NS	NS
SE(m) ±	0.0010	0.0011	0.0008
CD at 5%	-	-	-

Plate 9. Test weight of kasuri methi seed produced from the crop sown on 10th October and fertilized with 50:50:25 kg ha^{-1} NPK (D_1F_4) in the year of 2017-18

This might be due to that, an early sown crop would have produced the maximum branches per plant, leaves per plant. This might have helped in the production of more metabolites and consequently govern the accumulation of total solids which ultimately would have resulted into the production of maximum number of bolder seeds. Similarly, due to the favourable climatic conditions prevailing during period and more photosynthetic activity has resulted in better seed filling which resulted in the quality seeds in terms of higher test weight. The results of present investigation are in close agreement with the findings of Nandre (2006) and Singh *et al.* (2017[a]) in fenugreek.

4.12.2 Effect of fertilizer doses

The data furnished in Table 13 revealed that, fertilizer dose 50:50:25 kg/ha NPK (F_4) was recorded the maximum (1.172 g) test weight and it was at par with the treatment F_2 and F_3 (1.171 g). Whereas, the fertilizer dose 40:40:20 kg/ha NPK (F_1) was recorded minimum (1.168 g) test weight during the year 2016-17.

In the year 2017-18, the fertilizer dose 50:50:25 kg/ha (F_4) and 40:50:25 kg/ha NPK (F_2) were recorded the maximum (1.173 g) test weight and at par with F_3 (1.172). However, the fertilizer dose 40:40:20 kg/ha (F_1) was recorded significantly the minimum (1.171 g) test weight in kasuri methi crop.

The pooled data presented in Table 13 revealed that, the fertilizer dose 50:50:25 kg/ha (F_4) and 40:50:25 kg/ha NPK (F_2) were recorded the maximum (1.172 g) test weight. However, significantly the minimum (1.170 g) test weight was observed in fertilizer dose 40:40:20 kg/ha NPK (F_1). This might be due to the fact that, more supply of nutrients, combined with their availability freely and easily throughout of the crop growth phase might have produced the heavier seeds of kasuri methi. Similar results have been recorded an earlier workers like Nandre (2006) in fenugreek.

4.12.3 Interaction effect

The data presented in Appendix-V revealed that, the interaction effect of sowing dates and fertilizer doses to the test weight, during both the years of experimentations 2016-17 and 2017-18 were found to be non-significant in kasuri methi crop.

4.13 GERMINATION PER CENT (%)

The data regarding germination per cent during the years 2016-17 and 2017-18 as influenced due to individual effect of sowing dates and fertilizer doses are presented in Table 14 (a), (b) and depicted in Fig. 12.

4.13.1 Effect of sowing dates

The data furnished in Table 14 (a) showed that, an influence of sowing dates on germination per cent were found to be significant.

In the year 2016-17, the sowing date 10th October (D_1) was recorded significantly the maximum germination per cent (68.50 %) and it was found to the significantly superior over rest of the treatments. Whereas, minimum germination per cent (54.92 %) was measured in sowing date 20th January (D_6).

During the year 2017-18, significantly the maximum germination per cent (69.67 %) was observed in sowing date 10th October (D_1) and it was found) to be at par with the treatment D_2 (68.25). However, the sowing date 20th January (D_6) was recorded minimum (54.17 %) germination per cent. The better performance of early sown crop (10th October) was because of availability of congenial climatic conditions for the vegetative growth to kasuri methi crop. Due to which increase in the photosynthetic rate, accumulation of more photosynthetes resulting into better seed development and ultimately caused better seed germination and seedling vigour in the seeds produced in kasuri methi crop. Similar results have been obtained by earlier workers like Singh *et al.* (2017[a]) in fenugreek.

4.13.2 Effect of fertilizer doses

The data presented in Table 14 (a) noticed that, the influence of different fertilizer doses in respect of germination per cent were found to be significant during both the years of experimentations.

In the year 2016-17, the fertilizer dose 50:50:25 kg/ha NPK (F_4) measured the maximum (63.50 %) germination per cent and it was found to be at

Table 14 (a). Effect of dates of sowing and fertilizer doses on germination per cent of seed (%) in kasuri methi

Treatments	Germination per cent (%)		
	2016-17	2017-18	Pooled
Dates of sowing (D)			
D_1 - 10th October	68.50 (55.86)	69.67 (56.59)	69.08 (56.22)
D_2 - 30th October	67.67 (55.35)	68.25 (55.71)	67.96 (55.53)
D_3 - 20th November	64.42 (53.38)	64.42 (53.38)	64.42 (53.38)
D_4 - 10th December	60.58 (51.11)	61.17 (51.46)	60.88 (51.28)
D_5 - 30th December	57.08 (49.08)	56.33 (48.65)	56.71 (47.87)
D_6 - 20th January	54.92 (47.82)	54.17 (47.39)	54.54 (47.61)
'F' test	Sig.	Sig.	Sig.
SE(m) ±	0.16	0.32	0.20
CD at 5%	0.47	0.91	0.58
Fertilizer dose (F)			
F_1 - 40:40:20 kg ha^{-1} NPK	60.56 (51.13)	60.50 (51.11)	60.53 (51.12)
F_2 - 40:50:25 kg ha^{-1} NPK	62.89 (52.51)	63.00 (52.59)	62.94 (52.55)
F_3 - 50:40:20 kg ha^{-1} NPK	61.83 (51.89)	61.72 (51.38)	61.78 (51.86)

F_4 - 50:50:25 kg ha^{-1} NPK	63.50 (52.88)	64.11 (53.26)	63.81 (53.07)
'F' test	Sig.	Sig.	Sig.
SE(m) ±	0.13	0.26	0.17
CD at 5%	0.38	0.74	0.48
Interaction (D X F)			
'F' test	Sig.	Sig.	Sig.
SE(m) ±	0.33	0.64	0.41
CD at 5%	0.94	1.82	1.17

(Figures in parentheses are arc sin value transformation)

Table 14 (b). Interaction effect of dates of sowing and fertilizer doses on germination per cent of seed (%) in kasuri methi

Treatments	**Germination per cent (%)**		
	2016-17	**2017-18**	**Pooled**
D_1F_1	67.33 (55.14)	67.67 (55.35)	67.50 (55.24)
D_1F_2	69.00 (56.17)	70.67 (57.21)	69.83 (56.68)
D_1F_3	68.33 (55.76)	69.00 (56.17)	68.67 (55.96)
D_1F_4	69.33 (56.37)	71.33 (57.63)	70.33 (57.00)
D_2F_1	65.67 (54.13)	66.67 (54.74)	66.17 (54.43)
D_2F_2	67.67 (55.35)	68.33 (55.76)	68.00 (55.55)
D_2F_3	67.67 (55.35)	67.33 (55.14)	67.50 (55.24)
D_2F_4	69.67 (56.58)	70.67 (57.21)	70.17 (56.89)
D_3F_1	62.67 (52.34)	63.00 (52.54)	62.83 (52.44)
D_3F_2	65.33 (53.93)	65.00 (53.73)	65.17 (53.83)

D_3F_3	64.33 (53.33)	63.67 (52.93)	64.00 (53.13)
D_3F_4	65.33 (53.93)	66.00 (54.33)	65.67 (54.13)
D_4F_1	59.67 (50.57)	59.33 (50.38)	59.50 (50.48)
D_4F_2	61.33 (51.55)	61.67 (51.75)	61.50 (51.65)
D_4F_3	59.67 (50.57)	61.33 (51.55)	60.50 (51.06)
D_4F_4	61.67 (51.75)	62.33 (52.14)	62.00 (51.94)
D_5F_1	52.33 (46.34)	49.67 (44.81)	51.00 (54.57)
D_5F_2	58.67 (49.99)	59.00 (50.18)	58.83 (50.09)
D_5F_3	57.67 (49.41)	57.33 (49.22)	57.50 (49.31)
D_5F_4	59.67 (50.57)	59.33 (50.38)	59.50 (50.48)
D_6F_1	55.67 (48.25)	56.67 (48.83)	56.17 (48.54)
D_6F_2	55.33 (48.06)	53.33 (46.91)	54.33 (47.49)
D_6F_3	53.33 (46.91)	51.67 (45.96)	52.50 (46.43)
D_6F_4	55.33 (48.06)	55.00 (47.87)	55.17 (47.97)
'F' test	Sig.	Sig.	Sig.
SE(m) ±	0.33	0.64	0.41
CD at 5%	0.94	1.82	1.17

(Figures in parentheses are arc sin value transformation

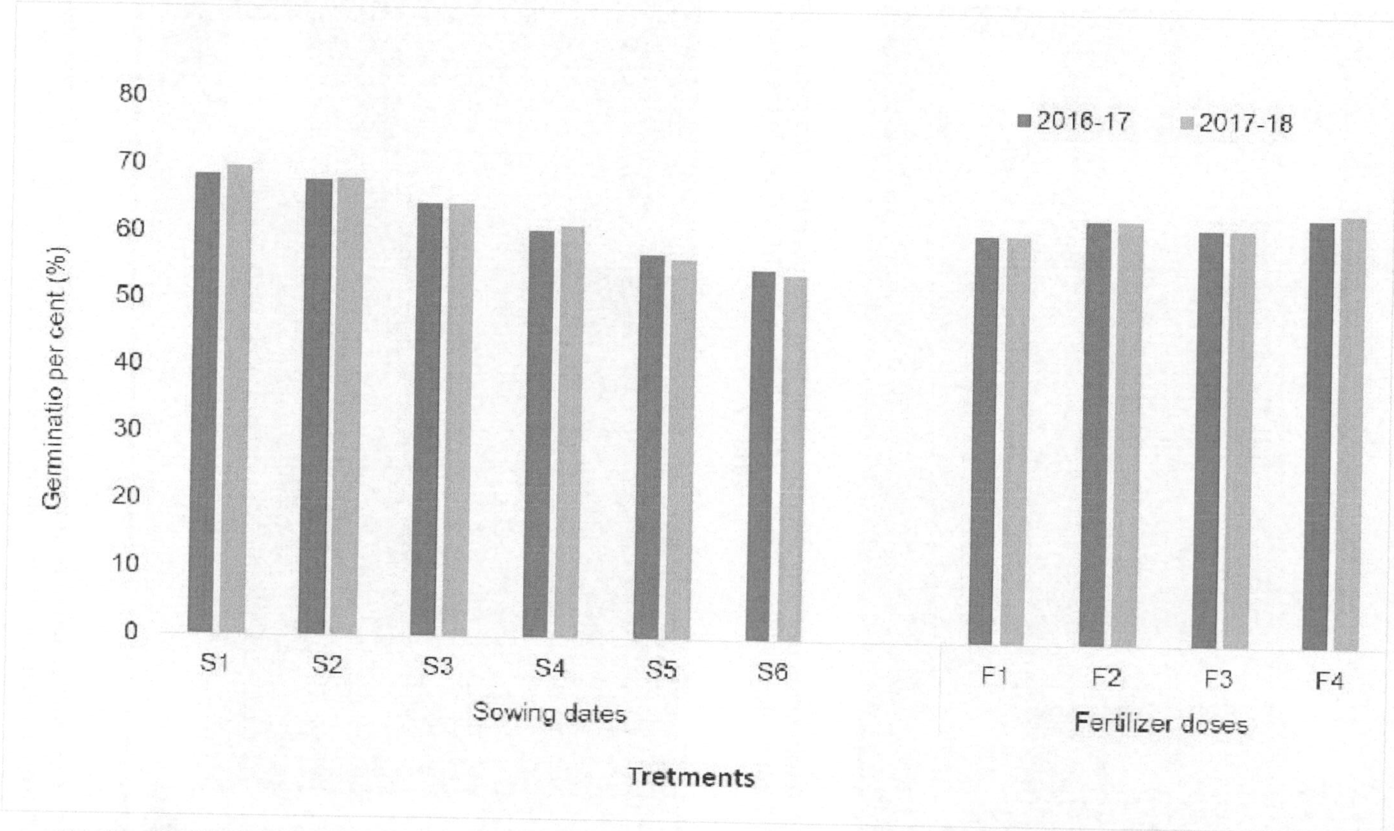

Fig.12. Effect of dates of sowing and fertilizer doses on germination per cent of seed (%) in kasuri methi

Plate 10. Germination per cent of kasuri methi seed produced from the crop sown on 10th October along with application of 50:50:25 kg ha^{-1} NPK (D_1F_4) in the year of 2016-17

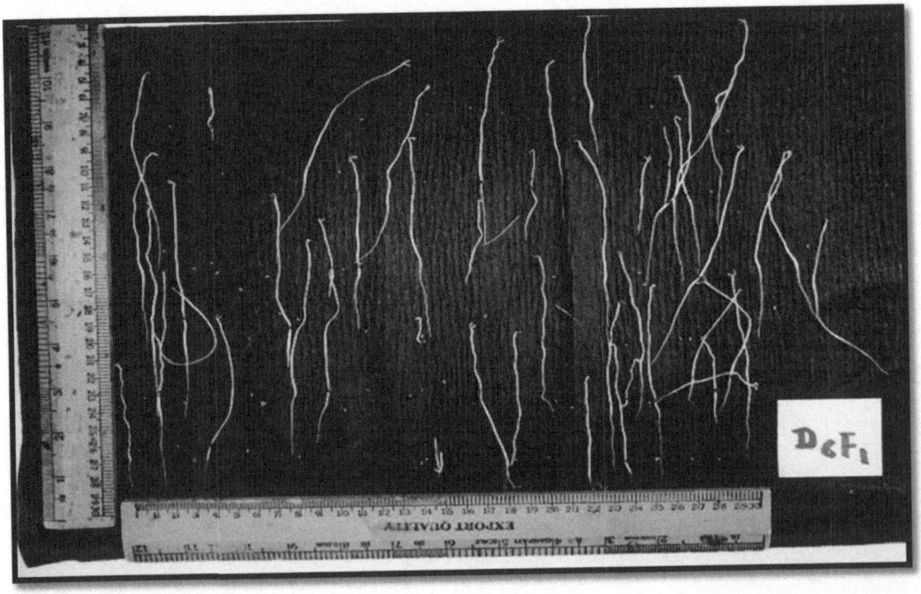

Plate 11. Germination per cent of kasuri methi seed produced from the crop sown on 20th January along with application of 40:40:20 kg ha^{-1} NPK (D_6F_1) in the year of 2016-17

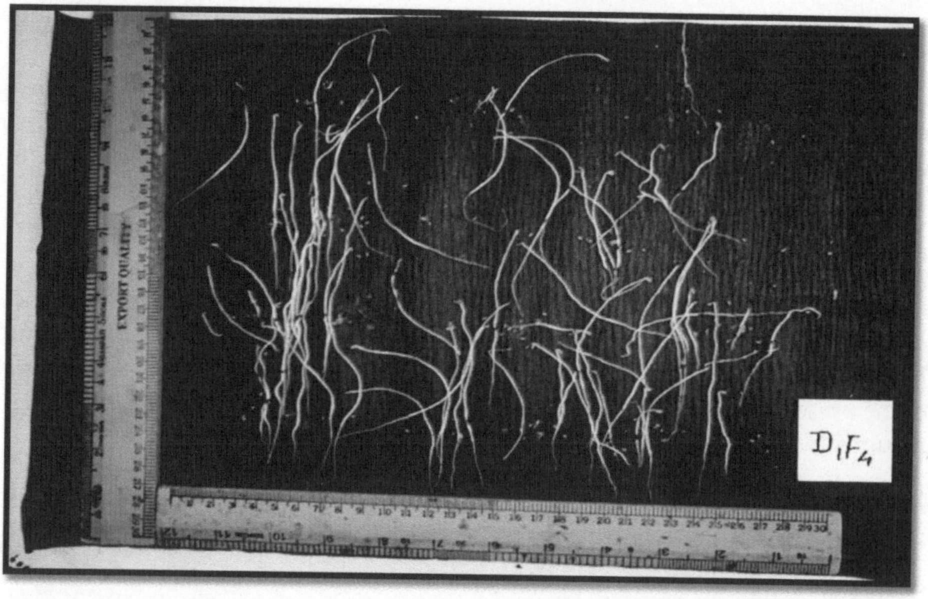

Plate 12. Germination per cent of kasuri methi seed produced from the crop sown on 10th October along with application of 50:50:25 kg ha^{-1} NPK (D_1F_4) in the year of 2017-18

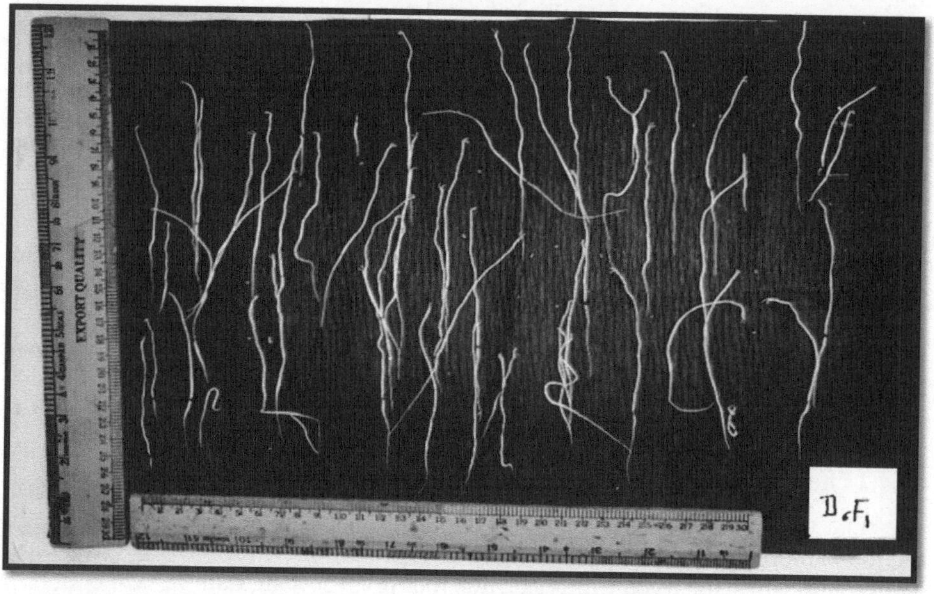

Plate 13. Germination per cent of kasuri methi seed produced from the crop sown on 20th January along with application of 40:40:20 kg ha^{-1} NPK (D_6F_1) in the year of 2017-18

Par with the treatment F_2 (62.89 %). However, minimum (60.56 %) germination per cent was observed in the fertilizer dose 40:40:20 kg/ha NPK (F_1).

During the year 2017-18, the maximum (64.11 %) germination per cent was noted in the fertilizer dose 50:50:25 kg/ha NPK (F_4) and it was found to be at par with the treatment F_2 (63.00 %). Whereas, the fertilizer dose 40:40:20 kg/ha NPK (F_1) was recorded the minimum (60.50 %) germination per cent in kasuri methi crop.

This might be due to fact that, an additional application of nutrients than the recommended dose of fertilizer might be effective in breaking the dormancy. As nitrogen is the integral part of the process of germination viz. protein synthesis, which might have enhanced the process of protein synthesis and which would have help to increase the germination percentage of kasuri methi seed. The results of the present investigation are in harmony with the findings of Nandre (2006) in fenugreek.

4.13.3 Interaction effect

The data presented in Table 14 (b) exhibited that, an interaction effect due to sowing dates and fertilizer doses regarding germination per cent of kasuri methi seeds during the year 2016-17 and 2017-18 were found to be statistically significant.

During the year 2016-17 and 2017-18, the treatment combination D_1F_4 was recorded significantly the maximum germination per cent (69.33 % and 71.33 %, respectively). Whereas, the treatment combination D_5F_1 was recorded significantly minimum germination per cent (52.33 % and 49.67 %, respectively).

The pooled results in respect of germination per cent (%) was significantly influenced due to the interaction effect of dates of sowing and fertilizer doses. However, the treatment combination D_1F_4 was produced maximum (70.33 %) germination per cent. Whereas, the minimum (51.00 %) germination per cent was recorded in the treatment combination D_5F_1. The results of the present investigations are supported by the findings of Nandre (2006) in fenugreek.

4.14 TOTAL CHLOROPHYLL CONTENT (MG/100 MG OF LEAVES)

The data towards chlorophyll content in the years of 2016-17 and 2017-18, as influenced by different sowing dates, fertilizer doses and their interactions in kasuri methi crop are presented in Table 15.

4.14.1 Effect of sowing dates

The data presented in Table 15 revealed that, the experimental year of 2016-17, the total chlorophyll content in kasuri methi leaves of the different sowing dates were significantly the maximum (3.02 mg/100mg) in sowing date 10th October (D_1). Whereas, significantly the minimum (2.67 mg/100mg) total chlorophyll content was recorded in sowing date 20th January (D_6).

During the year 2017-18, significantly the maximum (3.12 mg/100 mg) total chlorophyll content was recorded in sowing date 10th October (D_1) and which was found to be at par with the treatment D_2 (3.06 mg/100 mg). While, the minimum (2.73 mg/100 mg) total chlorophyll content was observed in sowing date 20th January (D_6).

As the time of sowing become late (20th January), the subsequent reductions were observed in the content of total chlorophyll. This might be due reduction in transpiration pull in late cooler months as compared to early sown (10th October) kasuri methi crop. Similar results have been obtained an earlier workers like Anitha *et al.* (2016) in fenugreek.

4.14.2 Effect of fertilizer doses

Effect of fertilizer doses during both the years of experimentations were found to be non-significant. In the year 2016-17, the fertilizer dose 50:50:25 kg/ha NPK (F_4) was measured the maximum (2.83 mg/100 mg) total chlorophyll content. However, minimum (2.73 mg/100 mg) total chlorophyll content was observed in kasuri methi crop fertilized with 40:40:20 kg/ha NPK (F_1).

Further, in the experimental year of 2017-18, the total chlorophyll content of kasuri methi leaves was noticed maximum (2.94 mg/100 mg) with the application of fertilizer dose 50:50:25 kg/ha NPK (F_4). Whereas,

Table 15. Effect of dates of sowing and fertilizer doses on total chlorophyll content (mg/100 mg of leaves at 60 DAS) in kasuri methi leaves

Treatments	Total chlorophyll content (mg/100 mg of leaves at 60 DAS)		
	2016-17	2017-18	Pooled
Dates of sowing (D)			
D_1 - 10th October	3.02	3.12	3.07
D_2 - 30th October	2.83	3.06	2.94
D_3 - 20th November	2.61	2.82	2.72
D_4 - 10th December	2.79	2.86	2.83
D_5 - 30th December	2.74	2.72	2.73
D_6 - 20th January	2.67	2.73	2.70
'F' test	Sig.	Sig.	Sig.
SE(m) ±	0.06	0.04	0.03
CD at 5%	0.16	0.10	0.10
Fertilizer dose (F)			
F_1 - 40:40:20 kg ha^{-1} NPK	2.73	2.82	2.78
F_2 - 40:50:25 kg ha^{-1} NPK	2.74	2.90	2.82
F_3 - 50:40:20 kg ha^{-1} NPK	2.80	2.88	2.84
F_4 - 50:50:25 kg ha^{-1} NPK	2.83	2.94	2.89

'F' test	NS	NS	NS
SE(m) ±	0.05	0.03	0.03
CD at 5%	-	-	-
Interaction (D X F)			
'F' test	NS	NS	NS
SE(m) ±	0.12	0.07	0.12
CD at 5%	-	-	-

the minimum (2.82 mg/100 mg) total chlorophyll content was recorded in the fertilizer dose 40:40:20 kg/ha NPK (F_1).

4.14.3 Interaction effect

The data presented in Appendix-VI indicated that, an interaction effect between sowing dates and fertilizer doses on chlorophyll content in kasuri methi were found to be statistically non-significant during both the years of experimentations 2016-17 and 2017-18 in kasuri methi crop.

4.15 TRIGONELLINE CONTENT (%)

The data in respect of trigonelline content (%) in kasuri methi seed, during the years 2016-17 and 2017-18 as influenced by sowing dates and fertilizer doses were found to be statistically non-significant and it is presented in Table 16.

4.15.1 Effect of sowing doses

The data furnished in Table 16 noticed that, the sowing date 10th October (D_1) was recorded maximum (0.25 %) trigonelline content in kasuri methi seed. However, the sowing date 20th November (D_3) was recorded the minimum (0.22 %) trigonelline content in kasuri methi seed in the year 2016-17.

During the year 2017-18, the sowing date 30th October (D_2) and 10th October (D_1) were recorded the maximum (0.25 %) trigonelline content in the seeds of kasuri methi. Whereas, the trigonelline content in seed of kasuri methi produced due to sowing date 30th December (D_5) was recorded minimum (0.23 %) in the present study. However, pooled data analysis said that, sowing date 10th October (D_1) was recorded the maximum (0.25 %) trigonelline content in the seeds of kasuri methi while, sowing date 30th December (D_5) was recorded minimum trigonelline content (0.23 %) in the present study.

4.15.2 Effect of fertilizer doses

Trigonelline content in kasuri methi seed as influenced due to the fertilizer doses were found to be non-significant, during both the years of experimentations which are presented in Table 16.

Table 16. Effect of dates of sowing and fertilizer doses on trigonelline content (%) in seeds of kasuri methi

Treatments	Trigonelline content (%)		
	2016-17	2017-18	Pooled
Dates of sowing (D)			
D_1 - 10th October	0.25 (2.86)	0.25 (2.84)	0.25 (2.85)
D_2 - 30th October	0.24 (2.78)	0.25 (2.86)	0.24 (2.82)
D_3 - 20th November	0.22 (2.69)	0.24 (2.83)	0.23 (2.76)
D_4 - 10th December	0.24 (2.79)	0.25 (2.87)	0.24 (2.83)
D_5 - 30th December	0.23 (2.76)	0.23 (2.77)	0.23 (2.77)
D_6 - 20th January	0.24 (2.78)	0.24 (2.83)	0.24 (2.81)
'F' test	NS	NS	NS
SE(m) ±	0.04	0.03	0.02
CD at 5%	-	-	-
Fertilizer dose (F)			
F_1 - 40:40:20 kg ha^{-1} NPK	0.24 (2.80)	0.25 (2.86)	0.25 (2.83)
F_2 - 40:50:25 kg ha^{-1} NPK	0.24 (2.82)	0.24 (2.82)	0.24 (2.82)
F_3 - 50:40:20 kg ha^{-1} NPK	0.23 (2.75)	0.24 (2.80)	0.24 (2.78)
F_4 - 50:50:25 kg ha^{-1} NPK	0.23 (2.73)	0.25 (2.85)	0.24 (2.79)
'F' test	NS	NS	NS
SE(m) ±	0.03	0.02	0.02
CD at 5%	-	-	-
Interaction (D X F)			
'F' test	NS	NS	NS
SE(m) ±	0.08	0.06	0.05
CD at 5%	-	-	-

(Figures in parentheses are arc sin value transformation)

The fertilizer dose 40:50:25 kg/ha NPK (F_2) and 40:40:20 kg/ha NPK (F_1) were recorded the maximum (0.24 %) trigonelline content in seed. However, the fertilizer dose 50:50:25 kg/ha NPK (F_4) and 50:40:20 kg/ha NPK (F_3) were recorded minimum (0.23 %) trigonelline content in seed, during the year 2016-17.

In the subsequent year 2017-18, the maximum (0.25 %) trigonelline content in seed was observed in treatment F_1 and F_4. Whereas, the treatment F_3 (50:40:20 kg/ha NPK) and F_2 (40:50:25 kg/ha NPK) were recorded minimum (0.24 %) trigonelline content in seeds of kasuri methi. However, pooled data analysis said that, fertilizer dose 40:40:20 kg/ha NPK (F_1) was recorded the maximum (0.25 %) trigonelline content in the seeds of kasuri methi while, fertilizer dose F_3, F_2 and F_4 were recorded minimum trigonelline content (0.24 %) in the present study.

4.15.3 Interaction effect

Interaction effect of sowing dates and fertilizer doses on the trigonelline content of seeds, during both the years of experimentations 2016-17 and 2017-18 were found to be statistically non-significant.

4.16 OLEORESIN CONTENT (%)

The data in respect of oleoresin content (%) in seed of kasuri methi crop during the years 2016-17 and 2017-18 as influenced by sowing dates and fertilizer doses were found to be non-significant and it is presented in Table 17.

4.16.1 Effect of sowing dates

The data furnished in Table 17 noticed that, the sowing date 20th November (D_3) was recorded the maximum (20.50 %) oleoresin content in seed. However, the sowing date 30th October (D_2) was recorded the minimum (20.00 %) oleoresin content in seed during the year 2016-17.

During the year 2017-18, the sowing date 10th October (D_1) was recorded the maximum (20.75 %) oleoresin content in seed. Whereas, the sowing date 30th October (D_2) was recorded minimum (20.04 %) oleoresin content in seed.

Table 17. Effect of dates of sowing and fertilizer doses on oleoresin content (%) in seed of kasuri methi

Treatments	Oleoresin content (%)		
	2016-17	2017-18	Pooled
Dates of sowing (D)			
D_1 - 10th October	20.33 (26.80)	20.75 (27.10)	20.54 (26.95)
D_2 - 30th October	20.00 (26.56)	20.04 (26.59)	20.02 (26.58)

D_3 - 20th November	20.50 (26.91)	20.46 (26.89)	20.48 (26.90)
D_4 - 10th December	20.21 (26.70)	20.17 (26.68)	20.19 (26.69)
D_5 - 30th December	20.13 (26.65)	20.54 (26.95)	20.33 (26.80)
D_6 - 20th January	20.29 (26.77)	20.50 (26.92)	20.40 (26.84)
'F' test	NS	NS	NS
SE(m) ±	0.24	0.14	0.14
CD at 5%	-	-	-
Fertilizer dose (F)			
F_1 - 40:40:20 kg ha^{-1} NPK	20.28 (26.76)	20.31 (26.78)	20.29 (26.77)
F_2 - 40:50:25 kg ha^{-1} NPK	20.67 (27.04)	20.56 (26.96)	20.61 (27.00)
F_3 - 50:40:20 kg ha^{-1} NPK	20.22 (26.72)	20.50 (26.92)	20.36 (26.82)
F_4 - 50:50:25 kg ha^{-1} NPK	19.81 (26.42)	20.28 (26.76)	20.04 (26.59)
'F' test	NS	NS	NS
SE(m) ±	0.20	0.11	0.11
CD at 5%	0.56	0.32	0.32
Interaction (D X F)			
'F' test	NS	NS	NS
SE(m) ±	0.49	0.28	0.27
CD at 5%	-	-	-

(Figures in parentheses are arc sin value transformation)

In pooled data analysis, noticed that, the sowing date 10th October (D_1) was recorded the maximum (20.54 %) oleoresin content in seed. However, the sowing date 30th October (D_2) was recorded the minimum (20.02 %) oleoresin content in seed.

4.16.2 Effect of fertilizer doses

Oleoresin content in seed as influenced due to the fertilizer doses were found to be non-significant during both the years of experimentations which are presented in Table 17.

The fertilizer dose 40:50:25 kg/ha NPK (F_2) was recorded the maximum (20.67 %) oleoresin content in seed. However, the fertilizer dose 50:50:25 kg/ha NPK (F_4) was recorded minimum (19.81 %) oleoresin content in seed, during the year 2016-17.

In subsequent year 2017-18, maximum (20.56 %) oleoresin content in seed was observed in the fertilizer dose 40:50:25 kg/ha NPK (F_2). Whereas, the fertilizer dose 50:50:25 kg/ha NPK (F_4) was recorded minimum (20.28 %) oleoresin content in seed.

The pooled data in respect of oleoresin content as influenced due to different fertilizer doses was found non-significant. But, numerically the maximum (20.61 %) oleoresin content in seed was observed in F_2. Whereas, minimum (20.04 %) observed in F_4 in the kasuri methi crop.

4.16.3 Interaction effect

Interaction effect of sowing dates and fertilizer doses on the oleoresin content in seed, during both the years of experimentations 2016-17 and 2017-18 were found to be statistically non-significant.

4.17 PROTEIN CONTENT (%)

The data in respect of protein content (%) in seeds of kasuri methi during the year 2016-17 and 2017-18 as influenced by sowing dates and fertilizer doses were found to be statistically non-significant and it is presented in Table 18.

Table 18. Effect of dates of sowing and fertilizer doses on protein content (%) in seed of kasuri methi

Treatments	Protein content (%)		
	2016-17	2017-18	Pooled
Dates of sowing (D)			
D_1 - 10th October	21.09 (27.34)	20.95 (27.24)	21.02 (27.29)
D_2 - 30th October	20.80 (27.13)	20.68 (27.05)	20.74 (27.09)
D_3 - 20th November	20.35 (26.82)	20.27 (26.76)	20.31 (26.79)
D_4 - 10th December	19.89 (26.48)	19.97 (26.54)	19.93 (26.51)
D_5 - 30th December	19.05 (25.87)	19.38 (26.12)	19.21 (26.00)
D_6 - 20th January	17.98 (25.09)	18.65 (25.58)	18.31 (25.34)
'F' test	NS	NS	NS
SE(m) ±	0.04	0.03	0.03

CD at 5%	0.12	0.08	0.08
Fertilizer dose (F)			
F_1 - 40:40:20 kg ha^{-1} NPK	19.59 (26.26)	19.76 (26.38)	19.67 (26.32)
F_2 - 40:50:25 kg ha^{-1} NPK	19.96 (26.53)	20.06 (26.60)	20.00 (26.56)
F_3 - 50:40:20 kg ha^{-1} NPK	19.79 (26.41)	19.92 (26.51)	19.87 (26.46)
F_4 - 50:50:25 kg ha^{-1} NPK	20.10 (26.63)	20.19 (26.70)	20.14 (26.66)
'F' test	NS	NS	NS
SE(m) ±	0.03	0.02	0.02
CD at 5%	0.10	0.07	0.07
Interaction (D X F)			
'F' test	NS	NS	NS
SE(m) ±	0.08	0.06	0.06
CD at 5%	-	-	-

(Figures in parentheses are arc sin value transformation)

4.17.1 Effect of sowing doses

The data presented in Table 18 noticed that, the sowing date 10th October (D_1) was recorded numerically maximum (21.09 %) protein content in seed. However, the sowing date 20th January (D_6) was recorded minimum (17.98 %) protein content in seed seeds of kasuri methi in the year 2016-17.

During the year 2017-18, the sowing date 10th October (D_1) was recorded numerically the maximum (20.95 %) protein content in seed. Whereas, the sowing date 20th January (D_6) was recorded numerically minimum (18.65 %) protein content in seed of kasuri methi.

Pooled data analysis said that, the sowing date 10th October (D_1) was recorded numerically the maximum (21.02 %) protein content in seed. Whereas, the minimum (18.31 %) protein content was observed in sowing date20th January (D_6).

4.17.2 Effect of fertilizer doses

Protein content in seed as influenced due to the fertilizer doses were also found to be non-significant during both the year of experimentation, which is presented in Table 18.

The fertilizer dose 50:50:25 kg/ha NPK (F_4) was recorded numerically the maximum (20.10 %) protein content in seed. However, the fertilizer dose 40:40:20

kg/ha NPK (F_1) was recorded minimum (19.59 %) protein content in seed of kasuri methi, during the year 2016-17.

In the year 2017-18, the maximum (20.19 %) protein content of seed was observed in fertilizer dose 50:50:25 kg/ha NPK (F_1) and the fertilizer dose 40:40:20 kg/ha NPK (F_1) was recorded minimum (19.76 %) protein content in seed.

As regards to pooled means of protein content, it is revealed that, numerically the maximum (20.14 %) protein content in seeds of kasuri methi was reported with F_4 (50:50:25 kg/ha NPK). However, numerically the minimum (19.67 %) protein content in seed of kasuri methi was observed with application of F_1 (40:40:20 kg/ha NPK)

4.17.3 Interaction effect

Interaction effect of sowing dates and fertilizer doses on the protein content of seed during both the years of experimentations 2016-17 and 2017-18 were found to be statistically non-significant in kasuri methi crop.

D. NITROGEN, PHOSPHORUS AND POTASSIUM STATUS

4.18 AVAILABLE NITROGEN, PHOSPHORUS AND POTASSIUM CONTENT IN SOIL

The data regarding the available nitrogen, phosphorus and potassium content in the soil, after harvesting of kasuri methi crop as influenced by the sowing dates and fertilizer doses are presented in Table 19 and depicted in Fig. 13 and 14.

4.18.1 Effect of sowing doses

The data regarding influence of the sowing dates on available nitrogen, phosphorus and potassium content in soil, after harvesting of kasuri methi crop were found to be statistically significant, during both the years of experimentations.

Among the sowing dates, significantly the maximum available nitrogen, phosphorus and potassium content (207.62, 14.33 and 397.25 kg ha^{-1}, respectively) were noticed in the soil, during the year 2016-17 with the sowing date 20th January (D_6). However, significantly the minimum values for these nutrients (195.85, 12.56 and 375.94 kg ha^{-1}, respectively) were noticed in the soil on which kasuri methi crop was sown crop was sown on dated 10th October (D_1).

During the year 2017-18, an available nitrogen, phosphorus and potassium content were significantly, the maximum (211.42, 16.58 and 410.77 kg ha^{-1}, respectively) in the soil, when the crop was sown on 20th January (D_6). Whereas, significantly the minimum values were recorded for these nutrients (198.27, 13.56 and 383.43 kg ha^{-1}, respectively), when the sowing date 10th October (D_1) of kasuri methi crop was undertaken.

4.18.2 Effect of fertilizer doses

The observations recorded regarding different fertilizer doses on available nitrogen, phosphorus and potassium content of the soil after

Table 19. Effect of dates of sowing and fertilizer doses on available nitrogen, phosphorus and potassium content in soil (kg ha^{-1})

Treatments	Available nutrient content in soil (kg ha^{-1})								
	Nitrogen			Phosphorus			Potassium		
	2016-17	2017-18	Pooled	2016-17	2017-18	Pooled	2016-17	2017-18	Pooled
Dates of sowing (D)									
D_1 - 10th October	195.85	198.27	197.06	12.56	13.56	13.06	375.94	383.43	379.68
D_2 - 30th October	199.22	202.24	200.73	13.21	14.42	13.82	380.26	388.39	384.32
D_3 - 20th November	201.70	205.05	203.37	13.57	14.96	14.26	386.49	394.34	390.42
D_4 - 10th December	204.09	207.14	205.61	13.85	15.64	14.75	392.72	400.39	396.55
D_5 - 30th December	206.31	209.75	208.03	14.16	16.24	15.20	395.55	405.83	400.69
D_6 - 20th January	207.62	211.42	209.52	14.33	16.58	15.46	397.25	410.77	404.01
'F' test	Sig.	Sig.	Sig.	Sig.	Sig.	Sig.	Sig.	Sig.	Sig.
SE(m) ±	0.47	0.40	0.28	0.10	0.16	0.11	0.45	0.42	0.38
CD at 5%	1.34	1.13	0.80	0.30	0.44	0.30	1.28	1.19	1.08
Fertilizer dose (F)									

F_1 - 40:40:20 kg ha^{-1} NPK	201.66	204.40	203.03	13.43	14.95	14.19	386.48	395.45	390.97
F_2 - 40:50:25 kg ha^{-1} NPK	202.55	205.82	204.19	13.68	15.33	14.51	388.60	397.73	393.16
F_3 - 50:40:20 kg ha^{-1} NPK	202.35	205.41	203.88	13.57	15.20	14.38	387.49	396.58	392.03
F_4 - 50:50:25 kg ha^{-1} NPK	203.31	206.94	205.13	13.78	15.45	14.62	389.57	399.01	394.29
'F' test	Sig.	Sig.	Sig.	Sig.	Sig.	Sig.	Sig.	Sig.	Sig.
SE(m) ±	0.38	0.33	0.23	0.08	0.13	0.09	0.37	0.34	0.31
CD at 5%	1.09	0.93	0.65	0.24	0.36	0.25	1.05	0.97	0.88
Interaction (D X F)									
'F' test	NS	NS	NS	NS	NS	NS	NS	NS	NS
SE(m) ±	0.94	0.80	0.56	0.21	0.31	0.21	0.90	0.84	0.76
CD at 5%	-	-	-	-	-	-	-	-	-

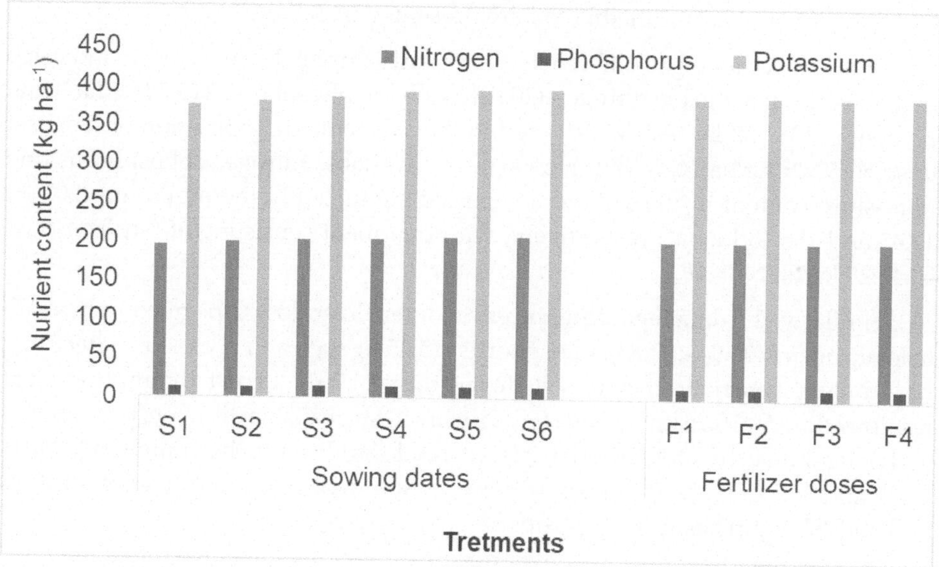

Fig. 13. Effect of dates of sowing and fertilizer doses on available nitrogen, phosphorus and potassium content in soil (kg ha^{-1}) during 2016-17

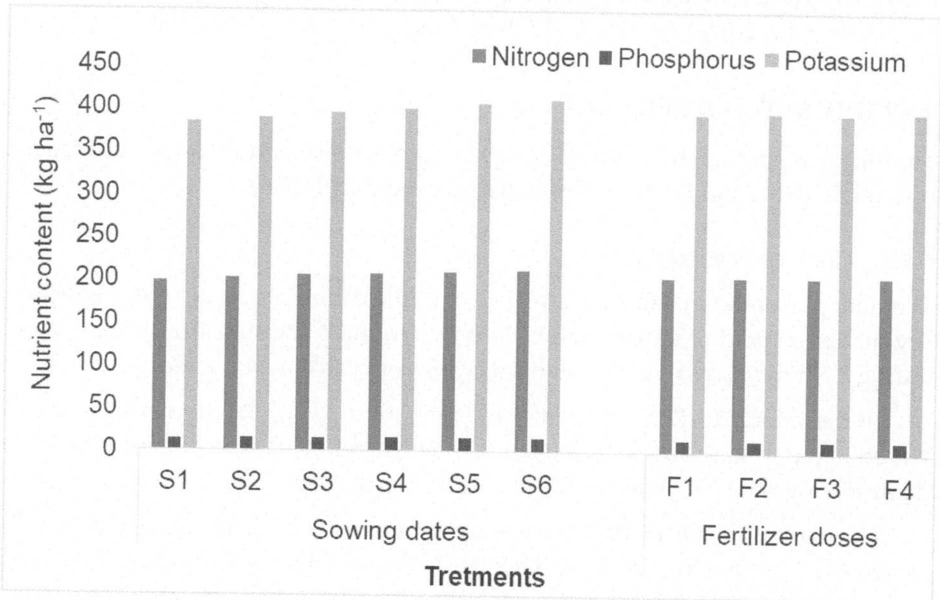

Fig. 14. Effect of dates of sowing and fertilizer doses on available nitrogen, phosphorus and potassium content in soil (kg ha^{-1}) during 2017-18

harvesting of kasuri methi crop are presented in Table 19.

The data pertaining to different fertilizer doses during the year 2016-17 indicated that, maximum available nitrogen (203. 31 kg ha^{-1}), phosphorus (13.78 kg ha^{-1}) and potassium (389.57 kg ha^{-1}) were noted in the soil with an application of fertilizer dose 50:50:25 kg/ha NPK (F_4). However, an available nitrogen, phosphorus and potassium content in the soil were recorded significantly the minimum (201.66, 13.43 and 386.48 kg ha^{-1}, respectively) with treatment consisting of fertilizer dose 40:40:20 kg/ha NPK (F_1).

Similarly, the data regarding the effect of fertilizer doses, the maximum available nitrogen (206.94 kg ha^{-1}), phosphorus (15.45 kg ha^{-1}) and potassium (399.01 kg ha^{-1}) content were noticed in the soil, during the year 2017-18 with the application of fertilizer dose 50:50:25 kg/ha NPK (F_4). However, an available nitrogen, phosphorus and potassium content in the soil were recorded significantly the minimum (204.40, 14.95 and 395.45 kg ha^{-1}, respectively) in the soil, which was fertilized @ 40:40:20 kg/ha NPK (F_1) in kasuri methi crop.

4.18.3 Interaction effect

All the interaction effect of sowing dates and fertilizer doses on available nitrogen, phosphorus and potassium content in the soil, after harvesting of kasuri methi crop, during the year 2016-17 and 2017-18 were found to be statistically non-significant.

4.19 DRY MATTER PER CENT (%)

The data in respect of dry matter during the years 2016-17 and 2017-18 as influenced by sowing dates and fertilizer doses are presented in Table 20.

4.19.1 Effect of sowing dates

The data presented in Table 20 revealed that, the different sowing dates effect on dry matter content of kasuri methi plant was found to be statistically significant, during both the years of experiments as well as in calculated pooled.

In the year 2016-17, significantly the maximum (15.05 %) dry matter was observed in the sowing date 10th October (D_1). However, it was found minimum (9.78 %) in sowing date 20th January (D_6) in kasuri methi crop.

During the year 2017-18, significantly the maximum (15.67 %) dry matter was observed in the sowing dates 10th October (D_1) and which was found to be at par with the treatment D_2 (15.24 %). Whereas, the sowing date 20th January (D_6) of kasuri methi crop recorded significantly the minimum (11.60 %) dry matter.

The pooled data furnished in Table 20 observed that, the sowing date 10th October (D_1) recorded significantly the maximum (15.36 %) dry matter. Whereas, sowing date 20th January (D_6) of kasuri methi crop was recorded minimum (10.69 %) dry matter.

There were decreasing trend in the dry matter (%) with delayed sowing dates in kasuri methi crop. This might be due to fact that, an early sown kasuri methi crop might having more prevailing of favourable climatic conditions to vegetative growth in terms of more plant height, branches per plant, leaves per plant and more yield attributes due to better photosynthetic efficiency and translocation of photosynthetes from source to sink. Similar results have been obtained an earlier worker like Singh *et al.* (2017[a]) in fenugreek.

4.19.2 Effect of fertilizer doses

The data furnished in Table 20 revealed that, an application of fertilizer dose 50:50:25 kg/ha NPK (F_4) to the kasuri methi crop recorded the maximum (12.89 %) dry matter. Whereas, the application of fertilizer dose 40:40:20 kg/ha NPK (F_1) was responsible for attaining the minimum (11.93 %) dry matter during 2016-17.

In the year 2017-18, an application of fertilizer dose 50:50:25 kg/ha NPK (F_4) was produced the maximum (14.14 %) dry matter. However, the application of fertilizer dose 40:40:20 kg/ha NPK (F_1) was exhibited significantly the minimum (13.33 %) dry matter.

The pooled data presented in Table 20 revealed that, an application of fertilizer 50:50:25 kg/ha NPK (F_4) was produced maximum (13.51 %) dry

Table 20. Effect of dates of sowing and fertilizer doses on total dry matter (%) in kasuri methi

Treatments	Total dry matter (%)		
	2016-17	2017-18	Pooled
Dates of sowing (D)			
D_1 - 10th October	15.05	15.67	15.36
D_2 - 30th October	14.10	15.24	14.67
D_3 - 20th November	13.14	14.10	13.62
D_4 - 10th December	11.83	13.36	12.60
D_5 - 30th December	10.80	12.63	11.72
D_6 - 20th January	9.78	11.60	10.69
'F' test	Sig.	Sig.	Sig.
SE(m) ±	0.27	0.17	0.13
CD at 5%	0.77	0.49	0.36

Fertilizer dose (F)			
F_1 - 40:40:20 kg ha^{-1} NPK	11.93	13.33	12.63
F_2 - 40:50:25 kg ha^{-1} NPK	12.68	13.91	13.29
F_3 - 50:40:20 kg ha^{-1} NPK	12.29	13.69	12.99
F_4 - 50:50:25 kg ha^{-1} NPK	12.89	14.14	13.51
'F' test	Sig.	Sig.	Sig.
SE(m) ±	0.22	0.14	0.10
CD at 5%	0.63	0.40	0.30
Interaction (D X F)			
'F' test	NS	NS	NS
SE(m) ±	0.54	0.35	0.25
CD at 5%	-	-	-

matter. However, significantly the minimum (12.63 %) dry matter was obtained, in the fertilizer dose 40:40:20 kg/ha NPK (F_1) applied to kasuri methi crop. This might be due to the fact that, an additional application of fertilizer dose to the kasuri methi crop reflects significant rise in growth parameters which were noticed under higher fertilizer level may be ascribed to greater uptake of nutrients by the plant favouring better cell division, cell elongation, amino acid and protein synthesis. And it might have produced more plant height, number of branches, leaves per plant which ultimately effects on more dry matter production as compare to low fertilizer levels. Similar results have been obtained an earlier worker like Kumar *et al.* (2015) in fenugreek.

4.19.3 Interaction effect

The data presented in Appendix-VII revealed that, dry matter accumulation in kasuri methi plant during the years 2016-17 and 2017-18 as influenced by interaction between sowing dates and fertilizer levels was found to be statistically non-significant.

4.20 NITROGEN, PHOSPHORUS AND POTASSIUM CONTENT IN KASURI METHI CROP

The data regarding an effect of the sowing dates and fertilizer doses on nitrogen, phosphorus and potassium content in kasuri methi plant during the years 2016-17 and 2017-18 are presented in Table 21 and depicted in Fig. 15 and 16.

4.20.1 Effect of sowing dates

The data indicated that, the sowing dates had showed significant effect on nitrogen, phosphorus and potassium content in the kasuri methi crop, during both the years of experimentations and for pooled results also.

Amongst the sowing dates, significantly the maximum nitrogen, phosphorus and potassium (1.330, 0.240 and 0.557 %, respectively) were found in the kasuri methi plant, during the year 2016-17 with the sowing date of 10th October (D_1). However, sowing date of 20th January (D_6) was recorded significantly the minimum nitrogen (1.234 %), phosphorus (0.179 %) and potassium (0.451 %) content, in the kasuri methi crop.

In the year 2017-18 significantly the maximum nitrogen (1.336 %), phosphorus (0.255) and potassium (0.573 %) content were found in the kasuri methi crop in the sowing date of 10th October (D_1). While, significantly the minimum nitrogen, phosphorus and potassium (1.249, 0.188 and 0.462 %, respectively) content in kasuri methi crop were observed, in the sowing date 20th January (D_1)

The pooled data exhibited the significant effects among the sowing dates on nutrient content of the kasuri methi crop. The sowing date 10th October (D_1) was expressed the maximum nitrogen (1.333 %), phosphorus (0.247 %) and potassium (0.565 %) content. However, significantly the minimum values of nitrogen (1.241 %), phosphorus (0.183 %) and potassium (0.456 %) content in kasuri methi crop were noted with the sowing date 20th January (D_6).

There were decreasing trend in respect of nitrogen, phosphorus and potassium content with delayed sowing in kasuri methi crop in the present investigation. This might be due to fact that, an early sown crop might having more prevailing of favourable climatic conditions and more availability of nitrogen, phosphorus and potassium, due to longer stage of crop growth which would have taken the maximum amounts of nutrients from the soil and exhibits more quantity of nutrients in the plant and seed. Similar results have been obtained by earlier workers like Nandre (2006) in fenugreek.

4.20.2 Effect of fertilizer doses

The data presented in Table 21 revealed that, the effect of fertilizer doses on nitrogen, phosphorus and potassium content in the kasuri methi crop were found to be significant for both the years of experimentations and for pooled results also.

The data pertaining to these attributes for the year 2016-17 indicated that, significantly the maximum nitrogen (1.291 %), phosphorus (0.215 %) and potassium (0.517 %) content in kasuri methi crop were

Table 21. Effect of dates of sowing and fertilizer doses on nitrogen, phosphorus and potassium content (%) in kasuri methi plant

Treatments	Nutrient content in kasuri methi plant (%)								
	Nitrogen			Phosphorus			Potassium		
	2016-17	2017-18	Pooled	2016-17	2017-18	Pooled	2016-17	2017-18	Pooled
Dates of sowing (D)									
D_1 - 10th October	1.330	1.336	1.333	0.240	0.255	0.247	0.557	0.573	0.565
D_2 - 30th October	1.314	1.329	1.322	0.231	0.248	0.239	0.544	0.556	0.550
D_3 - 20th November	1.292	1.308	1.300	0.217	0.236	0.227	0.533	0.530	0.532
D_4 - 10th December	1.271	1.283	1.277	0.202	0.222	0.212	0.502	0.504	0.503
D_5 - 30th December	1.253	1.259	1.256	0.190	0.210	0.200	0.467	0.478	0.473
D_6 - 20th January	1.234	1.249	1.241	0.179	0.188	0.183	0.451	0.462	0.456
'F' test	Sig.	Sig.	Sig.	Sig.	Sig.	Sig.	Sig.	Sig.	Sig.
SE(m) ±	0.005	0.005	0.004	0.002	0.002	0.002	0.004	0.004	0.003
CD at 5%	0.013	0.014	0.011	0.007	0.005	0.005	0.012	0.010	0.009
Fertilizer dose (F)									
F_1 - 40:40:20 kg ha^{-1} NPK	1.271	1.279	1.275	0.205	0.220	0.212	0.499	0.507	0.503

F_2 - 40:50:25 kg ha^{-1} NPK	1.286	1.297	1.288	0.212	0.229	0.220	0.514	0.521	0.518
F_3 - 50:40:20 kg ha^{-1} NPK	1.280	1.296	1.292	0.208	0.225	0.216	0.506	0.516	0.511
F_4 - 50:50:25 kg ha^{-1} NPK	1.291	1.303	1.297	0.215	0.232	0.223	0.517	0.525	0.521
'F' test	Sig.	Sig.	Sig.	Sig.	Sig.	Sig.	Sig.	Sig.	Sig.
SE(m) ±	0.004	0.004	0.003	0.002	0.001	0.001	0.003	0.003	0.003
CD at 5%	0.011	0.012	0.009	0.006	0.004	0.004	0.010	0.008	0.007
Interaction (D X F)									
'F' test	NS	NS	NS	NS	NS	NS	NS	NS	NS
SE(m) ±	0.009	0.010	0.008	0.005	0.003	0.004	0.008	0.007	0.006
CD at 5%	-	-	-	-	-	-	-	-	-

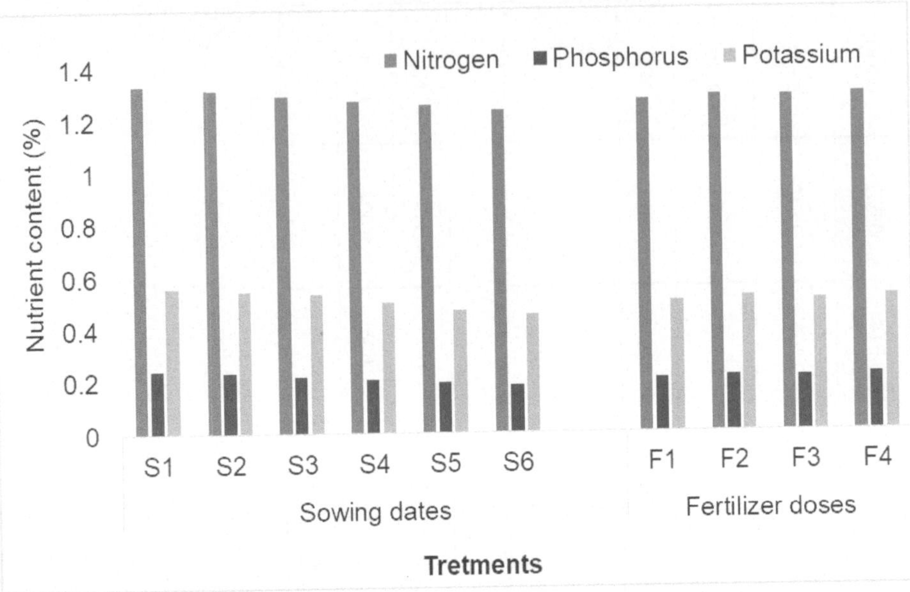

Fig. 15. Effect of dates of sowing and fertilizer doses on nitrogen, phosphorus and potassium content in kasuri methi plant (%) during 2016-17

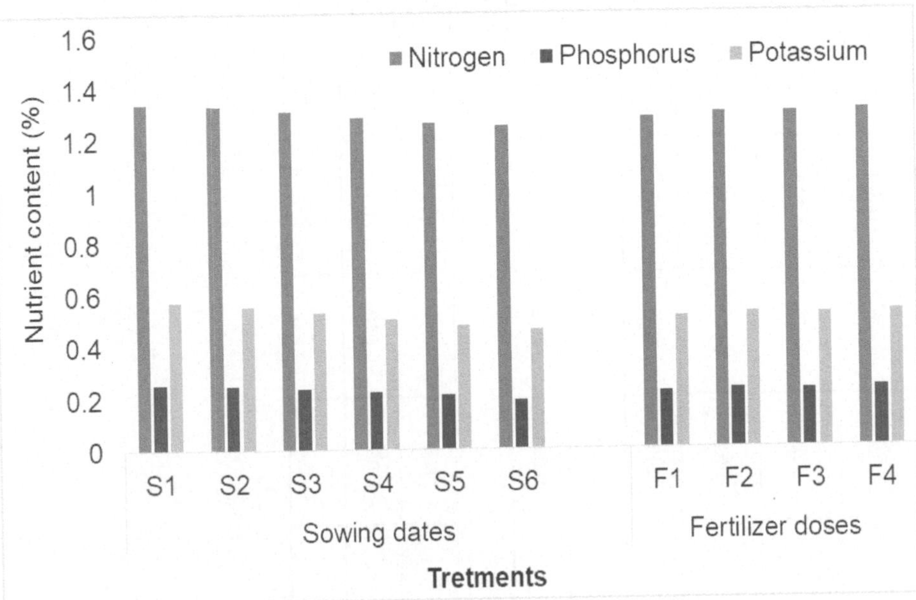

Fig. 16. Effect of dates of sowing and fertilizer doses on nitrogen, phosphorus and potassium content in kasuri methi plant (%) during 2017-18

recorded in fertilizer dose of 50:50:25 kg/ha (D_4). However, the treatment consisting of fertilizer dose of 40:40:20 kg/ha (D_1) was noted significantly the minimum nitrogen, phosphorus and potassium (1.271, 0.205 and 0.499 %, respectively) content in kasuri methi crop.

In the year 2017-18, nitrogen, phosphorus and potassium content in kasuri methi crop were observed significantly the maximum (1.303, 0.232 and 0.525 %, respectively) which was fertilized with 50:50:25 kg/ha NPK (F_4). However, the fertilizer dose 40:40:20 kg/ha (F_1) had shown significantly the minimum nutrient content (1.279, 0.220 and 0.507 %, respectively) in kasuri methi crop. The pooled data indicated that, nitrogen, phosphorus and potassium content in kasuri methi crop were significantly influenced due to fertilizer doses. The maximum nitrogen, phosphorus and potassium (1.297, 0.223 and 0.521 %, respectively) content in kasuri methi crop were recorded in the treatment (F_4) other in an application of fertilizer dose 50:50:25 kg/ha NPK. While, significantly the minimum nitrogen (1.275 %), phosphorus (0.212 %) and potassium (0.503 %) content were noted in the kasuri methi crop, which was fertilized with 40:40:20 kg/ha NPK (F_1). This might be due to the fact that, additional levels of nitrogen, phosphorus and potassium could be attributed to better availability of nutrients and their uptake throughout the crop growth period. Hence, the nutrient uptake was increased to a greater extent by plants. The results of the present investigation are supported by the findings of Babaleshwar and Shetty (2017) in kasuri methi.

4.20.3 Interaction effect

An interaction effects of sowing dates and fertilizer doses on nitrogen, phosphorus and potassium content in the kasuri methi plant were found to be statistically non-significant, during both the years of experimentations.

4.21 UPTAKE OF NITROGEN, PHOSPHORUS AND POTASSIUM BY KASURI METHI CROP

The data in respect of total uptake of nitrogen, phosphorus and potassium (kg ha^{-1}) by kasuri methi crop, as influenced due to sowing dates and fertilizer doses are presented in Table 22 and depicted in Fig. 17 and 18.

4.21.1 Effect of sowing dates

The effect of sowing dates on nitrogen, phosphorus and potassium uptake by kasuri methi crop was found to be significant (Table 22), during both the years of experimentations.

In the year 2016-17, significantly the maximum uptake of nitrogen, phosphorus and potassium (45.00, 8.11 and 18.82 kg ha^{-1}, respectively) were recorded by kasuri methi crop sown on dated 10th October (D_1). Whereas, the plant sown on dated 20th January (D_6) had recorded significantly the minimum uptake of nitrogen, phosphorus and potassium (25.73, 3.73 and 9.39 kg ha^{-1}, respectively) by kasuri methi crop.

During the year 2017-18, kasuri methi crop sown on 10th October (D_1) had recorded significantly the maximum uptake of nitrogen (46.79 kg ha^{-1}), phosphorus

(8.94 kg ha^{-1}) and potassium (20.06 kg ha^{-1}). However, significantly the minimum uptake of nitrogen, phosphorus and potassium (29.72, 4.47 and 11.01 kg ha^{-1}, respectively) were noted with sowing date 20th January (D_6).

The pooled data pertaining to the nutrients uptake had showed that, the treatment differences among the different sowing dates were found to be significant. The data indicated that, uptake of nitrogen, phosphorus and potassium were found significantly the maximum (45.90, 8.53 and 19.44 kg ha^{-1}, respectively) with the sowing date 10th October (D_1). However, significantly the minimum uptake of nitrogen (27.73 kg ha^{-1}), phosphorus (4.10 kg ha^{-1}) and potassium (10.20 kg ha^{-1}) were recorded under sowing date of 20th January (D_6).

4.21.2 Effect of fertilizer doses

The data presented in Table 22 had showed that, an effect of fertilizer doses on nitrogen, phosphorus and potassium uptake by kasuri methi crop was found to be statistically significant.

The data of the year 2016-17 noted that, the nitrogen,

Table 22. Effect of dates of sowing and fertilizer doses on total uptake of nitrogen, phosphorus and potassium by kasuri methi crop (kg ha^{-1})

Treatments	Total uptake of nutrients (kg ha^{-1})								
	Nitrogen			Phosphorus			Potassium		
	2016-17	2017-18	Pooled	2016-17	2017-18	Pooled	2016-17	2017-18	Pooled
Dates of sowing (D)									
D_1 - 10th October	45.00	46.79	45.90	8.11	8.94	8.53	18.82	20.06	19.44
D_2 - 30th October	42.17	45.28	43.72	7.41	8.44	7.93	17.44	18.95	18.19
D_3 - 20th November	37.75	40.15	38.95	6.35	7.24	6.80	15.57	16.28	15.93
D_4 - 10th December	33.31	36.07	34.69	5.31	6.25	5.78	13.15	14.17	13.66
D_5 - 30th December	28.73	33.57	31.15	4.36	5.59	4.97	10.72	12.74	11.73
D_6 - 20th January	25.73	29.72	27.73	3.73	4.47	4.10	9.39	11.01	10.20
'F' test	Sig.	Sig.	Sig.	Sig.	Sig.	Sig.	Sig.	Sig.	Sig.
SE(m) ±	0.65	0.47	0.36	0.13	0.08	0.08	0.20	0.22	0.13
CD at 5%	1.85	1.33	1.03	0.37	0.23	0.23	0.57	0.63	0.38

Fertilizer dose (F)									
F_1 - 40:40:20 kg ha^{-1} NPK	33.72	36.83	35.27	5.50	6.41	5.96	13.35	14.70	14.02
F_2 - 40:50:25 kg ha^{-1} NPK	36.18	39.12	37.65	6.03	6.96	6.49	14.58	15.81	15.19
F_3 - 50:40:20 kg ha^{-1} NPK	34.87	38.32	36.59	5.75	6.71	6.23	13.89	15.35	14.62
F_4 - 50:50:25 kg ha^{-1} NPK	37.03	40.12	38.57	6.23	7.21	6.72	14.92	16.27	15.60
'F' test	Sig.	Sig.	Sig.	Sig.	Sig.	Sig.	Sig.	Sig.	Sig.
SE(m) ±	0.53	0.38	0.29	0.10	0.07	0.07	0.16	0.18	0.11
CD at 5%	1.51	1.09	0.84	0.30	0.19	0.19	0.47	0.51	0.31
Interaction (D X F)									
'F' test	NS	NS	NS	NS	NS	NS	NS	NS	NS
SE(m) ±	1.30	0.94	0.72	0.26	0.16	0.16	0.40	0.44	0.27
CD at 5%	-	-	-	-	-	-	-	-	-

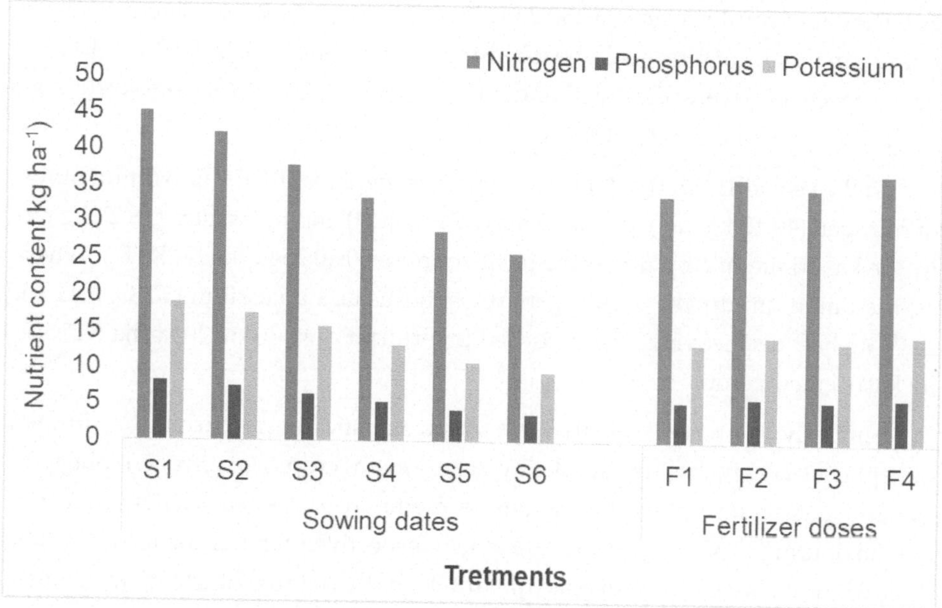

Fig. 17. Effect of dates of sowing and fertilizer doses on total uptake of nitrogen, phosphorus and potassium by kasuri methi crop (kg ha^{-1}) during 2016-17

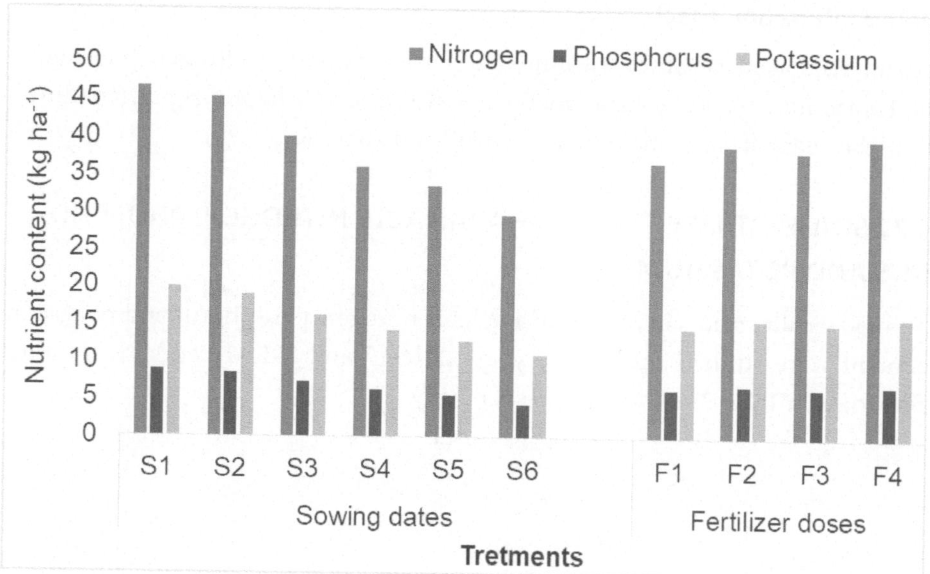

Fig. 18. Effect of dates of sowing and fertilizer doses on total uptake of nitrogen, phosphorus and potassium by kasuri methi crop (kg ha^{-1}) during 2017-18

phosphorus and potassium uptake by kasuri methi crop was observed to be significantly the maximum (37.03, 6.23 and 14.92 kg ha^{-1}, respectively) with an application of fertilizer dose 50:50:25 kg/ha NPK (F_4). However, significantly the minimum uptake of nitrogen, phosphorus and potassium (33.72, 5.50 and 13.35 kg ha^{-1}, respectively) were registered under the treatment of fertilizer dose 40:40:20 kg/ha NPK (F_1) by kasuri methi crop.

In the year 2017-18, the data indicated that, significantly the maximum uptake of nitrogen (40.12 kg ha^{-1}), phosphorus (7.21 kg ha^{-1}) and potassium (16.27 kg ha^{-1}) by the kasuri methi crop under the fertilizer dose 50:50:25 kg/ha NPK (F_4). Whereas, the minimum uptake of nitrogen, phosphorus and potassium (36.83, 6.41 and 14.70 kg ha^{-1}, respectively) was recorded in fertilizer dose 40450:20 kg/ha NPK (F_1) by kasuri methi crop.

Similarly, the pooled results were found statistically significant for nitrogen, phosphorus and potassium uptake by kasuri methi crop. The nitrogen, phosphorus and potassium uptake by the kasuri methi crop was recorded significantly the maximum (38.57, 6.72 and 15.60 kg ha^{-1}, respectively) under the fertilizer dose 50:50:25 kg/ha NPK (F_4). Whereas, the uptake of these three nutrients were noted to be minimum (35.27, 5.96 and 14.02 kg ha^{-1}, respectively) with the treatment pertaining to the fertilizer 40:40:20 kg/ha (F_1) application to the kasuri methi crop.

4.21.3 Interaction effect

An interaction effects of sowing dates and fertilizer doses on nitrogen, phosphorus and potassium uptake by kasuri methi crop were found to be non-significant during both the years of experimentations 2016-17 and 2017-18.

4.22 SOIL FERTILITY STATUS OF AVAILABLE NITROGEN, PHOSPHORUS AND POTASSIUM

The soil fertility status in respect of available nitrogen, phosphorus and potassium content were studied during the years 2016-17 and 2017-18 and the data are presented in Table 23 (a) and (b), respectively.

Table 23 (a). Soil fertility status for available nitrogen, phosphorus and potassium due to sowing dates and fertilizer levels in kasuri methi crop for the year 2016-17 (kg ha^{-1})

Treatments	Nitrogen (kg ha^{-1})			Phosphorus (kg ha^{-1})			Potassium (kg ha^{-1})		
	Initial status	Final status	Actual gain (+) or loss (-)	Initial status	Final status	Actual gain (+) or loss (-)	Initial status	Final status	Actual gain (+) or loss (-)
D_1F_1	190.34	195.70	5.36	10.24	12.12	1.88	368.72	374.65	5.93
D_1F_2	190.34	194.46	4.12	10.24	12.75	2.51	368.72	376.40	7.68
D_1F_3	190.34	196.20	5.86	10.24	12.48	2.24	368.72	375.33	6.61
D_1F_4	190.34	197.04	6.70	10.24	12.90	2.66	368.72	377.39	8.67
D_2F_1	190.34	198.09	7.75	10.24	13.06	2.82	368.72	378.53	9.81
D_2F_2	190.34	199.56	9.22	10.24	13.26	3.02	368.72	380.93	12.21
D_2F_3	190.34	199.16	8.82	10.24	13.19	2.95	368.72	379.50	10.78
D_2F_4	190.34	200.08	9.74	10.24	13.34	3.10	368.72	382.08	13.36
D_3F_1	190.34	201.08	10.74	10.24	13.45	3.21	368.72	383.95	15.23
D_3F_2	190.34	201.97	11.63	10.24	13.61	3.37	368.72	387.25	18.53
D_3F_3	190.34	201.37	11.03	10.24	13.54	3.30	368.72	385.53	16.81
D_3F_4	190.34	202.37	12.03	10.24	13.67	3.43	368.72	389.24	20.52
D_4F_1	190.34	203.17	12.83	10.24	13.72	3.48	368.72	391.38	22.66
D_4F_2	190.34	204.57	14.23	10.24	13.87	3.63	368.72	393.36	24.64

D_4F_3	190.34	203.77	13.43	10.24	13.78	3.54	368.72	392.24	23.52
D_4F_4	190.34	204.86	14.52	10.24	14.02	3.78	368.72	393.88	25.16
D_5F_1	190.34	205.69	15.35	10.24	14.05	3.81	368.72	394.70	25.98
D_5F_2	190.34	206.52	16.18	10.24	14.21	3.97	368.72	395.86	27.14
D_5F_3	190.34	205.90	15.56	10.24	14.11	3.87	368.72	395.21	26.49
D_5F_4	190.34	207.14	16.80	10.24	14.26	4.02	368.72	396.44	27.72
D_6F_1	190.34	206.21	15.87	10.24	14.16	3.92	368.72	395.70	26.98
D_6F_2	190.34	208.22	17.88	10.24	14.39	4.15	368.72	397.80	29.08
D_6F_3	190.34	207.68	17.34	10.24	14.30	4.06	368.72	397.09	28.37
D_6F_4	190.34	208.38	18.04	10.24	14.48	4.24	368.72	398.41	29.69

Table 23 (b). Soil fertility status for available nitrogen, phosphorus and potassium due to sowing dates and fertilizer levels in kasuri methi crop for the year 2017-18 (kg ha^{-1})

Treatments	Nitrogen (kg ha^{-1})			Phosphorus (kg ha^{-1})			Potassium (kg ha^{-1})		
	Initial status	Final status	Actual gain (+) or loss (-)	Initial status	Final status	Actual gain (+) or loss (-)	Initial status	Final status	Actual gain (+) or loss (-)
D_1F_1	195.70	196.53	0.83	12.12	13.25	1.13	374.65	381.63	6.98
D_1F_2	194.46	197.73	3.27	12.75	13.67	0.92	376.40	384.06	7.66
D_1F_3	196.20	198.59	2.40	12.48	13.53	1.06	375.33	383.21	7.87
D_1F_4	197.04	200.22	3.19	12.90	13.77	0.87	377.39	384.80	7.41
D_2F_1	198.09	201.26	3.17	13.06	14.05	0.99	378.53	386.44	7.91

D_2F_2	199.56	202.25	2.69	13.26	14.54	1.28	380.93	389.05	8.12
D_2F_3	199.16	201.97	2.81	13.19	14.45	1.26	379.50	387.43	7.93
D_2F_4	200.08	203.47	3.39	13.34	14.63	1.29	382.08	390.62	8.54
D_3F_1	201.08	204.12	3.04	13.45	14.77	1.32	383.95	392.43	8.48
D_3F_2	201.97	205.34	3.37	13.61	14.97	1.36	387.25	395.05	7.80
D_3F_3	201.37	204.72	3.36	13.54	14.90	1.36	385.53	393.43	7.90
D_3F_4	202.37	205.99	3.62	13.67	15.18	1.52	389.24	396.46	7.21
D_4F_1	203.17	206.19	3.02	13.72	15.47	1.75	391.38	398.37	6.99
D_4F_2	204.57	207.30	2.73	13.87	15.71	1.84	393.36	400.89	7.53
D_4F_3	203.77	206.76	2.99	13.78	15.62	1.84	392.24	399.84	7.60
D_4F_4	204.86	208.29	3.43	14.02	15.78	1.76	393.88	402.46	8.58
D_5F_1	205.69	208.86	3.17	14.05	15.90	1.85	394.70	404.22	9.52
D_5F_2	206.52	210.09	3.57	14.21	16.43	2.22	395.86	406.32	10.46
D_5F_3	205.90	209.15	3.24	14.11	16.08	1.97	395.21	405.53	10.32
D_5F_4	207.14	210.89	3.75	14.26	16.55	2.29	396.44	407.27	10.83
D_6F_1	206.21	209.42	3.21	14.16	16.25	2.08	395.70	409.60	13.89
D_6F_2	208.22	212.22	4.00	14.39	16.67	2.28	397.80	410.99	13.19
D_6F_3	207.68	211.26	3.58	14.30	16.61	2.31	397.09	410.02	12.93
D_6F_4	208.38	212.78	4.40	14.48	16.80	2.32	398.41	412.45	14.04

The data presented in Table 23 (a) revealed that, during the year 2016-17, the maximum gain of nitrogen was noted under treatment combination D_6F_4 (18.04 kg ha^{-1}) and it was followed by the treatment combination D_6F_2 (17.88 kg ha^{-1}), D_6F_3 (17.34 kg ha^{-1}) and D_5F_4 (16.80 kg ha^{-1}). However, the remaining treatment combinations have noted the gain of available nitrogen in the range of 4.12 kg ha^{-1} (D_1F_2) to 16.18 kg ha^{-1} (D_5F_2) in ascending order. No loss of the nitrogen had been recorded under any of treatment combinations.

Similarly, a meager gain in phosphorus was observed and it was in the range of 1.88 kg ha^{-1} (D_1F_1) to 4.24 kg ha^{-1} (D_6F_4) in ascending order. After completion of the experiment, there was no loss of phosphorus, under any of the treatment combinations.

There was an overall gain in the status of an available potassium due to the sowing dates and nutrient management treatment combinations. Maximum gain of potassium was noted under the treatment combination D_6F_4 (29.69 kg ha^{-1}) and it was followed by the treatment combination D_6F_2 (29.08 kg ha^{-1}).

The data for the year 2017-18 in respect of an available nitrogen, phosphorus and potassium status due to kasuri methi crop are presented in the Table 23 (b) and it exhibited that, the maximum gain of available nitrogen was noticed under the treatment combination D_6F_4 (4.40 kg ha^{-1}). However, the remaining treatment combination noted the gain of nitrogen in the range of 0.83 kg ha^{-1} (D_1F_1) to 4.00 kg ha^{-1} (D_6F_2) in ascending order. After completion of the experiment, there was no loss of nitrogen, under the any of treatment combinations.

After completion of the experiment, a meager gain in phosphorus was noticed and it was in the range of 0.87 kg ha^{-1} (D_1F_4) to 2.32 kg ha^{-1} (D_6F_4) in ascending order.

There was an overall gain in the status of the available potassium due to the treatment combinations of sowing dates and nutrient management. The maximum gain of potassium was noted under the treatment combination D_6F_4 (14.04 kg ha^{-1}).

E. ECONOMIC STUDIES OF TREATMENTS

4.23 GROSS AND NET MONETARY RETURN

The data presented in Table 24 (a), (b) and depicted in Fig. 19 and 20 had showed significant influence on gross and net monetary return due to the sowing dates and fertilizer doses, during the two years of experimentations as well as in the pooled results too.

4.23.1 Effect of sowing dates

The data pertaining to gross monetary return and net monetary return during the first year (2016-17) envisages that, both the benefits were the maximum (Rs. 350471/-

and Rs. 274231/- ha^{-1}, respectively) with the sowing date 10th October (D_1). Whereas, the gross monetary and net monetary return were found significantly the minimum (Rs. 182132/- and Rs. 105892/- ha^{-1}, respectively) by the sowing date 20th January (D_6).

Similarly, the gross and net monetary return, during the year 2017-18 were found to be maximum (Rs. 355765/- and Rs. 279525/- ha^{-1}, respectively) with the sowing date 10th October (D_1). However, these values were significantly the minimum (Rs. 178607/- and Rs. 102367/- ha^{-1}, respectively) with the sowing date 20th January (D_6).

The pooled data resulted that, the gross and net monetary return from kasuri methi crop were the maximum (Rs. 353200/- and Rs. 276960/- ha^{-1}, respectively) when the crop was sown on dated 10th October (D_1). However, the sowing date 20th January (D_6) gave significantly the minimum gross monetary return (Rs. 180367/- ha^{-1}) and net monetary return (Rs. 104127/- ha^{-1}).

This might be due the fact that, an early sown (10th October) kasuri methi crop has produced the maximum seed yield and finally resulted into more gross and net monetary return in present investigation.

4.23.2 Effect of fertilizer doses

The effect of fertilizer doses on gross and net monetary return for both the years of experimentations and for pooled results were found to be statistically significant. The data (2016-17) indicated that, the higher

Table 24 (a). Effect of dates of sowing and fertilizer doses on gross and net monetary return in kasuri methi (Rs. ha^{-1})

Treatments	Gross monetary return (Rs. ha^{-1})			Net monetary return (Rs. ha^{-1})		
	2016-17	2017-18	Pooled	2016-17	2017-18	Pooled
Dates of sowing (D)						
D_1 - 10th October	350471	355765	353200	274231	279525	276960
D_2 - 30th October	343129	346100	344600	266889	269860	268360
D_3 - 20th November	293326	287911	290633	217086	211671	214393
D_4 - 10th December	260240	233900	247133	184000	157660	170893
D_5 - 30th December	196434	224294	210333	120194	148054	134093
D_6 - 20th January	182132	178607	180367	105892	102367	104127
'F' test	Sig.	Sig.	Sig.	Sig.	Sig.	Sig.
SE(m) ±	4742.03	4202.61	3381.26	4742.03	4202.61	3381.26
CD at 5%	13498.97	11963.41	9625.32	13498.97	11963.41	9625.32
Fertilizer dose (F)						
F_1 - 40:40:20 kg ha^{-1} NPK	261535	259565	260578	185644	183674	184687
F_2 - 40:50:25 kg ha^{-1} NPK	274528	274674	274578	198069	198215	198119
F_3 - 50:40:20 kg ha^{-1} NPK	264830	265999	265467	188809	189978	189446
F_4 - 50:50:25 kg ha^{-1} NPK	282929	284146	283556	206340	207557	206967
'F' test	Sig.	Sig.	Sig.	Sig.	Sig.	Sig.
SE(m) ±	3871.85	3431.41	2760.79	3871.85	3431.41	2760.79
CD at 5%	11021.86	9768.09	7859.04	11021.86	9768.09	7859.04
Interaction (D X F)						
'F' test	Sig.	Sig.	Sig.	Sig.	Sig.	Sig.
SE(m) ±	9484.06	8405.22	6762.53	9484.06	8405.22	6762.53
CD at 5%	26997.93	23926.83	19250.64	26997.93	23926.83	19250.64

Table 24 (b). Interaction effect of dates of sowing and fertilizer doses on gross and net monetary return in kasuri methi (Rs ha^{-1})

Treatments	Gross monetary return (Rs. ha^{-1})			Gross monetary return (Rs. ha^{-1})		
	2016-17	2017-18	Pooled	2016-17	2017-18	Pooled
D_1F_1	343152	345400	344400	267261	269509	268509

D_1F_2	352911	361851	357333	276452	285392	280874
D_1F_3	348705	352657	350800	272684	276636	274779
D_1F_4	357114	363152	360267	280525	286563	283678
D_2F_1	337552	339962	338667	261661	264071	262776
D_2F_2	345851	349584	347600	269392	273125	271141
D_2F_3	341495	343041	342400	265474	267020	266379
D_2F_4	347619	351813	349733	271030	275224	273144
D_3F_1	275695	274489	275067	199804	198598	199176
D_3F_2	300159	291735	296000	223700	215276	219541
D_3F_3	287340	284397	285867	211319	208376	209846
D_3F_4	310111	301022	305600	233522	224433	229011
D_4F_1	267794	237810	252933	191903	161919	177042
D_4F_2	261140	228105	244667	184681	151646	168208
D_4F_3	243829	205987	224933	167808	129966	148912
D_4F_4	268200	263698	266000	191611	187109	189411
D_5F_1	145997	202095	174000	70106	126204	98109
D_5F_2	211940	223863	217867	135481	147404	141408
D_5F_3	204343	244314	224400	128322	168293	148379
D_5F_4	223457	226902	225067	146868	150313	148478
D_6F_1	199019	157635	178400	123128	81744	102509
D_6F_2	175165	192905	184000	98706	116446	107541
D_6F_3	163270	165600	164400	87249	89579	88379
D_6F_4	191073	198289	194667	114484	121700	118078
'F' test	Sig.	Sig.	Sig.	Sig.	Sig.	Sig.
SE(m) ±	9484.06	8405.22	6762.53	9484.06	8405.22	6762.53
CD at 5%	26997.93	23926.83	19250.64	26997.93	23926.83	19250.64

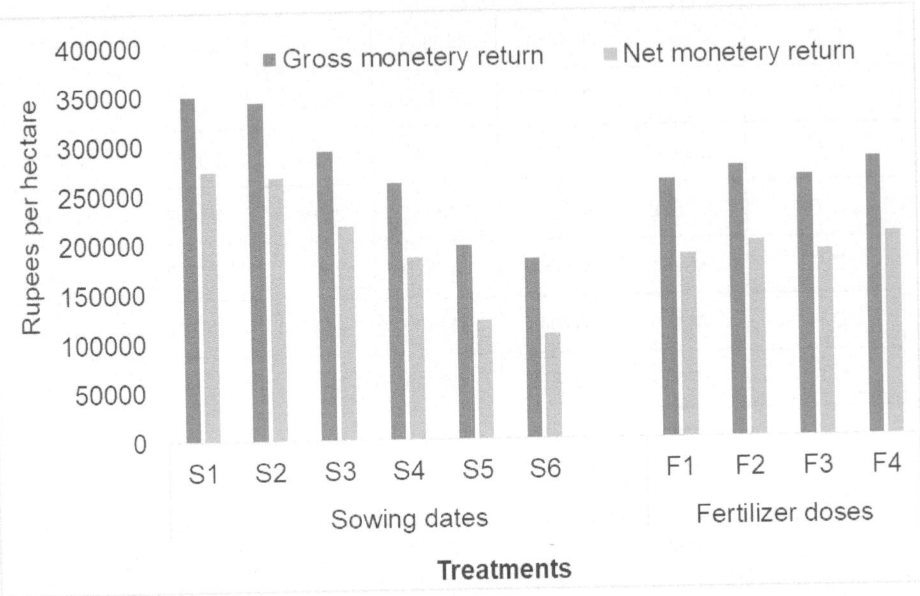

Fig. 19. Effect of dates of sowing and fertilizer doses on gross and net monetary return in kasuri methi (Rs. ha^{-1}) during 2016-17

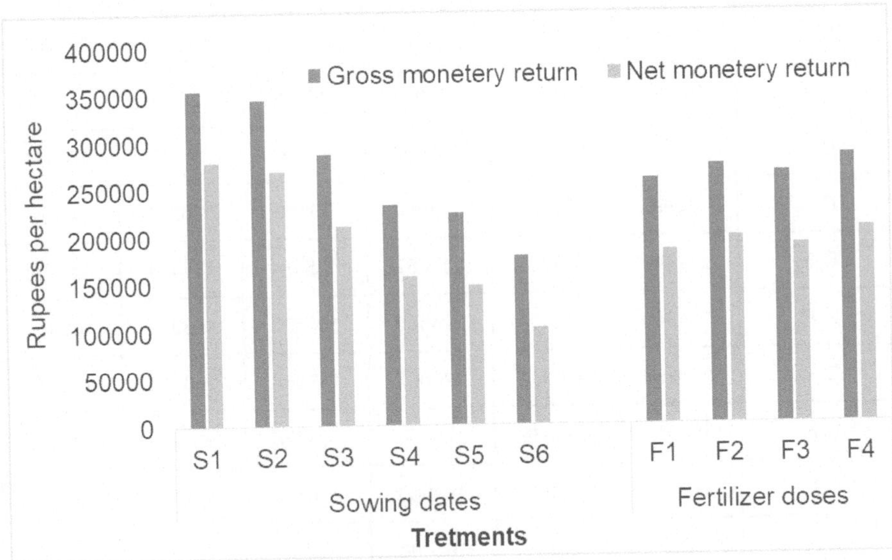

Fig. 20. Effect of dates of sowing and fertilizer doses on gross and net monetary return in kasuri methi (Rs. ha^{-1}) during 2017-18

fertilizer dose was recorded significantly the maximum gross and net monetary return over the application of lower fertilizer dose to kasuri methi crop. However, significantly the maximum gross (Rs. 282929/- ha^{-1}) and net monetary return (Rs. 206340/- ha^{-1}) were obtained due to the application of fertilizer dose 50:50:25 kg/

ha NPK (F_4). The values for this were significantly the minimum (Rs. 261535/- and Rs. 185644/- ha^{-1}, respectively) with an application of the fertilizer dose 40:40:20 kg/ha NPK (F_1).

Similarly, the gross and net monetary return during the year 2017-18 were found to be significantly the maximum (Rs. 284146/- and Rs. 207557/- ha^{-1}, respectively) with an application of treatment F_4. Whereas, these values were significantly the minimum (Rs. 259565/- and Rs. 183674/- ha^{-1}, respectively) with low application of fertilizer dose 40:40:20 kg/ha NPK (F_1) in kasuri methi crop.

The pooled data exhibited that, significantly the maximum gross monetary return (Rs. 283556/- ha^{-1}) and net monetary return (Rs. 206967/- ha^{-1}) were obtained from an application of treatment F_4. Significantly the minimum gross (Rs. 260578/- ha^{-1}) and net monetary return (Rs. 184687/- ha^{-1}) were gained from an application of fertilizer dose 40:40:20 kg/ha NPK (F_1) in kasuri methi crop. This might be due the fact that, an application of high fertilizer dose i.e. 50:50:25 kg ha^{-1} NPK had produced maximum seed yield of kasuri methi and finally resulted into more gross and net monetary return from the kasuri methi crop in the present study.

4.23.3 Interaction effect

Interaction effects of sowing dates and fertilizer doses on gross and net monetary returns during 2016-17 and 2017-18 were found to be statistically significant and presented in Table 24(b).

In the year 2016-17, the treatment combination D_1F_4 recorded significantly the maximum gross (Rs. 357114/- ha^{-1}) and net monetary return (Rs. 280525/- ha^{-1}). However, significantly the minimum gross (Rs. 145997/- ha^{-1}) and net monetary return (Rs. 70106/- ha^{-1}) were found in treatment combination (D_5F_1) wherein, sowing of kasuri methi crop was undertaken on dated 30th December along with application of 40:40:20 kg/ha NPK

During the year 2017-18, significantly the maximum gross (Rs. 363152/- ha^{-1}) and net monetary return (Rs. 286563/- ha^{-1}) were found in the treatment combination D_1F_4 in which sowing of kasuri methi crop was done on dated 10th October and fertilized with 50:50:25 kg/ha NPK. However, significantly the minimum gross (Rs. 157635/- ha^{-1}) and net monetary return (Rs. 81744/- ha^{-1}) were found in treatment of combination D_6F_1.

The pooled data resulted that, however, numerically the treatment combination D_1F_4 produced the maximum gross and net monetary return (Rs. 360267/- and Rs. 283678/- ha^{-1}, respectively). Whereas, numerically the minimum values (Rs. 164400/- and Rs. 88379/- ha^{-1}, respectively) in this regard were recorded in D_6F_3.

4.24 B:C RATIO

The data pertaining to B:C ratio as influenced due to sowing dates and fertilizer doses during both the years of experimentations are presented in Table 25 (a), (b) and depicted in Fig. 21.

4.24.1 Effect of sowing dates

During the year 2016-17, the maximum B:C ratio was recorded in the treatment D_1 (3.60). Similarly, during the year 2017-18 maximum B:C ratio was recorded in the treatment D_1 (3.67). Whereas, the treatment D_1 (10th October) exhibited maximum (3.63) pooled analysis B:C ratio in the present study.

4.24.2 Effect of fertilizer doses

During the year 2016-17 and 2017-18, the maximum B:C ratio (2.69 and 2.71, respectively) were recorded in the kasuri methi crop receiving the higher fertilizer dose 50:50:25 kg/ha NPK (F_4).

As far as the pooled mean values of B:C ratio were concerned, the kasuri methi crop receiving the fertilizer dose 50:50:25 kg/ha NPK (F_4) was recoded the maximum value (2.70).

Table 25 (a). Effect of dates of sowing and fertilizer doses on cost : benefit ratio in kasuri methi

Treatments	Cost : benefit ratio		
	2016-17	2017-18	Pooled
Dates of sowing (D)			
D_1 - 10th October	3.60	3.67	3.63
D_2 - 30th October	3.50	3.54	3.52
D_3 - 20th November	2.85	2.78	2.81
D_4 - 10th December	2.41	2.07	2.24
D_5 - 30th December	1.58	1.94	1.76
D_6 - 20th January	1.39	1.34	1.37
'F' test	-	-	-
SE(m) ±	-	-	-
CD at 5%	-	-	-
Fertilizer dose (F)			
F_1 - 40:40:20 kg ha^{-1} NPK	2.45	2.42	2.43
F_2 - 40:50:25 kg ha^{-1} NPK	2.59	2.59	2.59
F_3 - 50:40:20 kg ha^{-1} NPK	2.48	2.50	2.49
F_4 - 50:50:25 kg ha^{-1} NPK	2.69	2.71	2.70
'F' test	-	-	-
SE(m) ±	-	-	-
CD at 5%	-	-	-

Table 25 (b). Interaction effect of dates of sowing and fertilizer doses on cost : benefit ratio in kasuri methi

Treatments	Cost : benefit ratio		
	2016-17	2017-18	Pooled
D_1F_1	3.52	3.55	3.54
D_1F_2	3.62	3.73	3.67
D_1F_3	3.59	3.64	3.61
D_1F_4	3.66	3.74	3.70
D_2F_1	3.45	3.48	3.46
D_2F_2	3.52	3.57	3.55
D_2F_3	3.49	3.51	3.50
D_2F_4	3.54	3.59	3.57
D_3F_1	2.63	2.62	2.62
D_3F_2	2.93	2.82	2.87
D_3F_3	2.78	2.74	2.76
D_3F_4	3.05	2.93	2.99
D_4F_1	2.53	2.13	2.33
D_4F_2	2.42	1.98	2.20
D_4F_3	2.21	1.71	1.96
D_4F_4	2.50	2.44	2.47
D_5F_1	0.92	1.66	1.29
D_5F_2	1.77	1.93	1.85
D_5F_3	1.69	2.21	1.95
D_5F_4	1.92	1.96	1.94
D_6F_1	1.62	1.08	1.35
D_6F_2	1.29	1.52	1.41
D_6F_3	1.15	1.18	1.16
D_6F_4	1.49	1.59	1.54

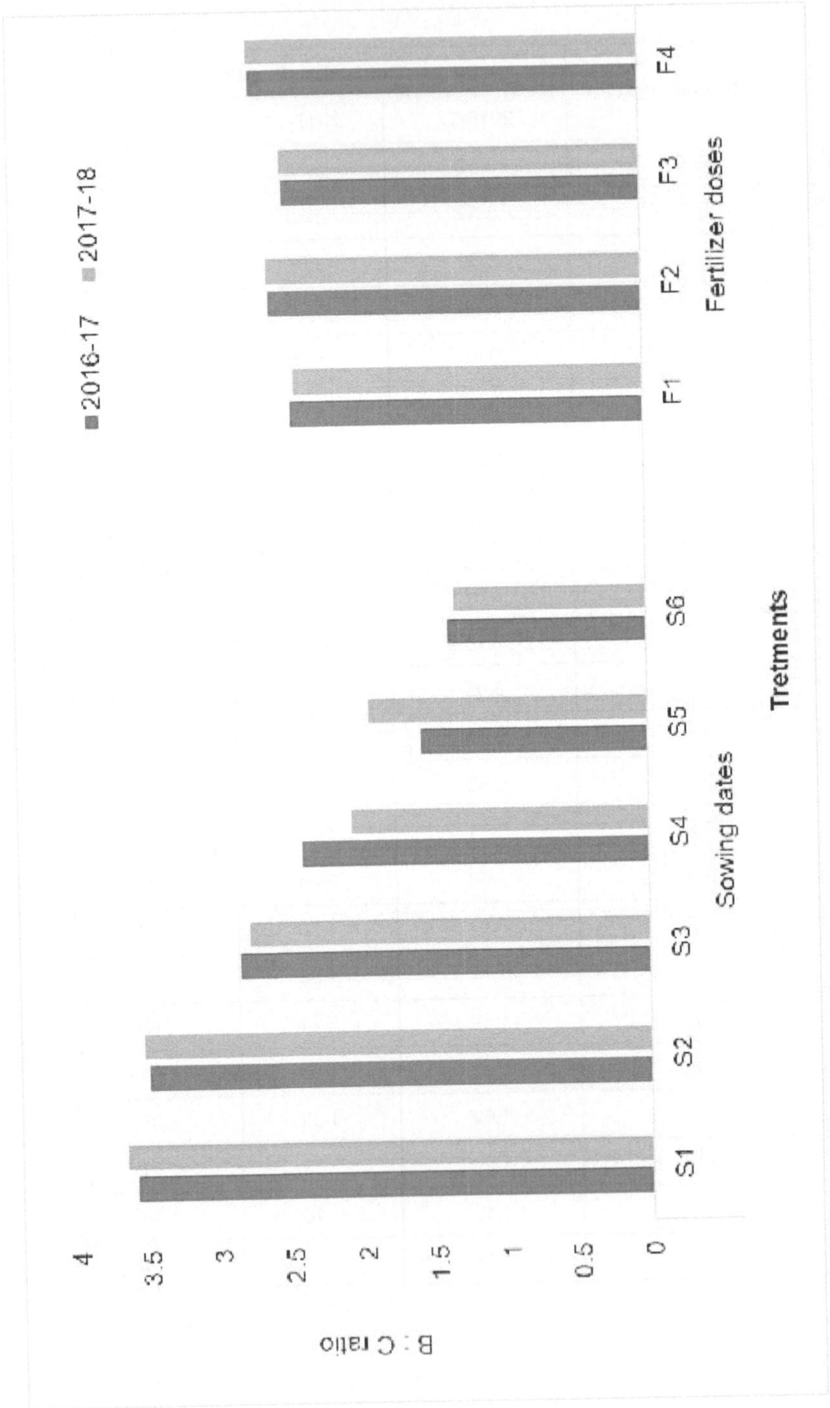

Fig. 21. Effect of dates of sowing and fertilizer doses on cost : benefit ratio in kasuri methi

4.24.3 Interaction effect

The data regarding B:C ratio, as influenced due to sowing dates and fertilizer doses during both the years of experimentations are presented in Table 25 (b).

During both the years, 2016-17 and 2017-18, the maximum B:C ratio was obtained with the treatment combination D_1F_4 (3.66 and 3.74, respectively).

While in pooled data, the maximum B:C ratio was recorded under the treatment combination D_1F_4 (3.70).

As regards to produced mean, it is noted that, the treatment combination D_1F_4 (kasuri methi crop sown on 10th October and fertilized with 50:50:25 kg/ha NPK) were recorded the maximum B:C ratio (3.70). However, the treatment combination D_6F_3 i.e. sowing of kasuri methi crop on 20th January along with fertilized @ 50:50:25 kg/ha NPK was recorded minimum mean B:C ratio (1.16) in the kasuri methi crop.

4.25 Incidence of Powdery mildew

Incidence of the powdery mildew on kasuri methi crop was observed below economical threshold level during both the years of experimentations in 2016-17 and 2017-18.

5 CHAPTER

SUMMARY AND CONCLUSIONS

The present investigation entitled "Effect of sowing time and fertilizer levels on seed yield of kasuri methi (*Trigonella corniculata* L.)" which was carried out at the Chilli and Vegetable Research Unit (CVRU), Dr. Panjabrao Deshmukh Krishi Vidyapeeth, Akola during *rabi* seasons of the years 2016-17 and 2017-18 with the following objectives:

To study the different sowing time and fertilizer levels for seed production of kasuri methi.

To find out best treatment combination of sowing time and fertilizer level for seed production of kasuri methi.

The experiment was laid out in factorial randomized block design with three replications. There were two factors of an experiment, first being the sowing dates consisted of six levels viz. D_1 (10th October), D_2 (30th October), D_3 (20th November), D_4 (10th December), D_5 (30th December), D_6 (20th January) and four levels of fertilizer doses viz. F_1 (40:40:20 kg ha^{-1}), F_2 (40:50:25 kg ha^{-1}), F_3 (50:40:20 kg ha^{-1}) and F_4 (50:50:25 kg ha^{-1} NPK). The experiment was conducted with 24 (twenty four) treatment combinations, to study the effect of sowing time and fertilizer levels on vegetative growth, flowering, seed yield, seed quality and yield attributing characters, along with the nutrients content in soil and kasuri methi plants. The results obtained in these respect are summarized below.

5.1 VEGETATIVE GROWTH

Maximum vegetative growth in terms of plant height, branches per plant and number of leaves per plant were recorded by the kasuri methi crop sown on dated 10th October (D_1) and fertilized with 50:50:25 kg ha^{-1} NPK (F_4). While, the sowing date 20th January (D_6) and fertilized with fertilizer level of 40:40:20 kg ha^{-1} NPK (F_1) were recorded minimum values in these respect. The treatment combination D_1F_4

(kasuri methi crop sown on dated 10th October and fertilized with 50:50:25 kg ha^{-1} NPK) was recorded the maximum values for all the above characters.

5.2 YIELD ATTRIBUTES AND YIELD

The kasuri methi crop sown on 30th October (D_2) and fertilized with 40:40:20 kg ha^{-1} NPK (F_1) were got the maximum days to flower initiation and 50 per cent flowering during both the years of experimentations. While, the crop sown on 10th October (D_1) and fertilized with 50:50:25 kg ha^{-1} NPK (F_4) were recorded the maximum florets per plant during 2016-17 (34.44, and 27.93, respectively); during 2017-18 (36.47 and 27.70, respectively), the maximum pods per florets (23.54 and 19.34, respectively), seeds per pod (5.63 and 4.52, respectively), seed yield per plant (5.32 and 4.26 g, respectively), seed yield per plot (370.77 and 297.71 g, respectively) and seed yield per hectare (8.83 and 7.09 q, respectively) were reported.

However, the kasuri methi crop sown on 20th January (D_6) and fertilized with 50:50:25 kg ha^{-1} NPK (F_4) got the minimum days to flower initiation and 50 per cent flowering during both the years of experimentations. While, the crop sown on 20th January (D_6) and fertilized with 40:40:20 kg ha^{-1} NPK (F_1) were recorded the minimum florets per plant during 2016-17 (17.45, and 24.77, respectively); for the year 2017-18 (16.04 and 25.02, respectively), minimum pods per florets (12.98 and 17.60, respectively), seeds per pod (2.92 and 4.03, respectively), seed yield per plant (2.69 and 3.90 g, respectively), seed yield per plot (189.39 and 273.71 g, respectively) and seed yield per hectare (4.51 and 6.51 q, respectively) were reported.

5.3 SEED QUALITY PARAMETERS

The kasuri methi crop sown on 10th October (D_1) and fertilized with 50:50:25 kg ha^{-1} NPK (F_4) were produced the maximum test weight of seed (1.176 and 1.172 g, respectively), total chlorophyll content (3.07 and 2.89 mg/100 g, respectively), protein content (21.02 and 20.04 %, respectively) and the maximum germination percentage of seed (68.50 and 63.50 %, respectively), during the year 2016-17 and germination per cent (69.67 and 64.11 %, respectively) during the year 2017-18 was observed. While, the maximum trigonelline content (0.25 %) was reported by the crop sown on 10th October and fertilized with 40:40:20 kg ha^{-1} NPK (F_1). However, the maximum oleoresin content (20.54 and 20.61 %, respectively) was reported by the crop sown on 10th October and fertilized with 40:50:25 kg ha^{-1} NPK (F_2).

However, the crop sown on 20th January (D_6) and fertilized with 40:40:20 kg ha^{-1} NPK (F_1) were produced the minimum test weight of seed (1.166 and 1.170 g, respectively), total chlorophyll content (2.70 and 2.78 mg/100 g, respectively), protein content (18.31 and 20.29 %, respectively). The minimum germination percentage of seed (54.92 and 60.56 %, respectively in the year 2016-17) and (54.17 and 60.50 %, respectively in the year 2017-18) was observed.

While, the minimum trigonelline content (0.23 and 0.24, respectively %) was reported by the crop sown on 20th November (D_3) and fertilized with 50:40:20 kg ha^{-1} NPK (F_3). However, the minimum oleoresin content (20.02 and 20.04 %, respectively) was reported by the crop sown on 30th October (D_2) and fertilized with 50:50:25 kg ha^{-1} NPK (F_4).

5.4 CHEMICAL STUDY

The kasuri methi crop sown on 20th January (D_6) and fertilized with 50:50:25 kg ha^{-1} NPK (F_4) were noticed the maximum availability of nitrogen (207.62 and 203.31 kg ha^{-1}, respectively), phosphorus (14.33 and 13.78 kg ha^{-1}, respectively) and potassium in the soil (397.25 and 389.57 kg ha^{-1}) during the year 2016-17. Further, the maximum available nitrogen (211.42 and 206.94 kg ha^{-1}, respectively), available phosphorus (16.58 and 15.45 kg ha^{-1}, respectively) and available potassium in the soil (410.77 and 399.01 kg ha^{-1}) were observed during the year 2017-18.

However, the crop sown on 10th October (D_1) and fertilized with 40:40:20 kg ha^{-1} NPK (F_1) were recorded the minimum available nitrogen (195.85 and 201.66, respectively), phosphorus (12.56 and 13.43 the ha^{-1}, respectively) and potassium in the soil (375.94 and 386.48 kg ha^{-1}) during the year 2016-17.

Whereas, during the year 2017-18, the minimum available nitrogen (198.27 and 204.40 kg ha^{-1}, respectively), phosphorus (13.56 and 14.95 Kg ha^{-1}, respectively) and potassium in the soil (383.43 and 395.45 kg ha^{-1}) were observed in the present study.

The pooled analysis revealed that, the crop sown on 10th October (D_1) and fertilized with 50:50:25 kg ha^{-1} NPK (F_4) were recorded the maximum nitrogen (1.333 and 1.197 %, respectively), phosphorus (0.247 and 0.223 %, respectively) and potassium content in plant (0.565 and 0.521 %, respectively). However, crop sown on 20th January (D_6) and fertilized with 40:40:20 kg ha^{-1} NPK (F_1) were recorded minimum nitrogen (1.241 and 1.275 %, respectively), phosphorus (0.183 and 0.212 %, respectively) and potassium content in plant (0.465 and 0.503 %, respectively) in the kasuri methi crop.

Considering the pooled analysis towards uptake of nutrients by kasuri methi plant, when the kasuri methi crop sown on 10th October (D_1) and fertilized with 50:50:25 kg ha^{-1} NPK (F_4) were recorded the maximum uptake of nitrogen (50.49, 46.42 and 45.66 kg ha^{-1}, respectively), phosphorus (8.53 and 6.72 kg ha^{-1}, respectively) and potassium (19.44 and 15.68 kg ha^{-1}, respectively) by kasuri methi plant. However, crop sown on 20th January (D_6) and fertilized with 40:40:20 kg ha^{-1} NPK (F_1) were recorded minimum uptake of nitrogen (27.73 and 35.27 kg ha^{-1}, respectively), phosphorus (4.10 and 5.96 Kg ha^{-1}, respectively) and potassium (10.20 and 14.02 kg ha^{-1}, respectively) by kasuri methi.

The treatment combination D_6F_4 (crop sown on 20th January and fertilized with 50:50:25 kg ha^{-1} NPK) registered the maximum values for available nitrogen, phosphorus and potassium content in the soil.

While, treatment combination D_1F_4 (crop sown on 10th October and fertilized with 50:50:25 kg ha^{-1} NPK) was recorded the maximum numerical values for nitrogen, phosphorus and potassium content of plant and uptake by plant. Whereas, minimum values for content in plant and uptake by plant were noticed in the treatment combination D_6F_1 (crop sown on 20th January and fertilized with 40:40:20 kg ha^{-1} NPK).

5.5 ECONOMICS

The kasuri methi crop sown on 10th October (D_1) and fertilized with 50:50:25 kg ha^{-1} NPK (F_4) have recorded significantly the maximum gross monetary return (Rs. 353200/- and 283556/- ha^{-1}, respectively), net monetary return (Rs. 276960/- and 206967/- ha^{-1}, respectively). However, the crop sown on 20th January (D_6) and fertilized with 40:40:20 kg ha^{-1} NPK (F_1) were recorded minimum gross monetary return (Rs. 180367/- and 260578/- ha^{-1}, respectively), net monetary return (Rs. 104127/- and 184687/- ha^{-1}, respectively) as an individual level effect. While, the treatment combination D_1F_4 (kasuri methi crop sown on 10th October and fertilized with 50:50:25 kg/ha NPK) was recorded the maximum gross and net monetary return (Rs. 360267/- and 283678/- ha^{-1}, respectively).

The mean data regarding cost : benefit ratio revealed that, the kasuri methi sown on 10th October (D_1) and fertilized with 50:50:25 kg ha^{-1} NPK (F_4) were recorded the maximum cost : benefit ratio (3.63 and 2.70, respectively) during both the years of experimentation 2016-17 and 2017-18. However, during both the years, the minimum cost : benefit ratio (1.37 and 2.43, respectively) were recorded under the crop sown on 20th January (D_6) and fertilized with 40:40:20 kg ha^{-1} NPK (F_1).

As far as the pooled means of the cost : benefit ratio is concerned, the treatment combination D_1F_4 (sown on 10th October and fertilized with 50:50:25 kg ha^{-1} NPK) was recorded the maximum cost : benefit ratio (3.70). Whereas, the minimum cost : benefit ratio (1.16) was registered under treatment combination D_6F_3.

CONCLUSIONS

From the present investigation the following conclusions could be drawn.

Maximum vegetative growth of the kasuri methi plants viz., plant height, number of leaves and number of branches per plant were produced by the crop sown on 10th October (D_1) with the application of fertilizer dose 50:50:25 kg ha^{-1} NPK (F_4). Whereas, the maximum days required for flower initiation and 50 per cent flowering were observed in the crop sown on 30th October (D_2) with the application of 40:40:20 kg ha^{-1} NPK (F_1). Similarly, the maximum number of florets per plant, pods per floret and seeds per pod were noticed in the crop sown on 10th October (D_1) with the application of fertilizer dose 50:50:25 kg ha^{-1} NPK (F_4).

In respect of seed yield per plant, per plot and per hectare of kasuri methi, there were produced significantly the maximum (5.42 g, 378.14 g and 9.0 q, respectively) from crop kasuri methi sown on 10th October along with the application of 50:50:25 kg ha^{-1} NPK (D_1F_4). As far as the quality parameters of kasuri methi were concerned, viz., test weight, germination per cent, total chlorophyll content and trigonelline content in kasuri methi, they were found to be the maximum when sown on 10th October (D_1) and fertilized with 50:50:25 kg ha^{-1} NPK (F_4).

Maximum availability of nitrogen, phosphorus and potassium content in the soil were found when the kasuri methi crop was sown on 20th January (D_6) with the application of 50:50:25 kg ha^{-1} NPK (F_4). The NPK content in kasuri methi plant and uptake of NPK by the crop were recorded the maximum when the crop was sown on 10th October (D_1) and fertilized with 50:50:25 kg ha^{-1} NPK (F_4). Furthermore, the maximum gross (Rs. 360267/- ha^{-1}), net monetary return (Rs. 283678/- ha^{-1}) and cost benefit ratio (3.70) was obtained from the kasuri methi crop sown on 10th October with the application of 50:50:25 kg ha^{-1} NPK (D_1F_4).

The results inferred from the present investigation are suggestive and need to be studied further for obtaining appropriate recommendation.

6
CHAPTER

LITERATURE CITED

Abou-Shleel. S. M. 2014. Effect of air temperature on growth, yield and active ingredients of fenugreek (*Trigonella foenum graecum*). Nature and Science: **12**(9): 50-54.

Ali Haidar, Gohar Ayub, Ehsan Elahi Muhammad Sahab, Saeed Ahmed and Nazeer Ahmed. 2015. Response of coriander (*Coriandrum sativum* L.) to different nitrogen levels and sowing dates. *Asian J. Agri Biol.* **3**(4):140-144.

Anitha B., M. Lakshmi Narayana Reddy, A. V. D. Dorajee Rao, T. S. K. K. Kiran Patro and D. R. Salomi Suneetha. 2016. Effect of sowing dates on yield and quality of fenugreek. *Plant Archives*. **16**(1), PP. 479-484.

Anonymous. 1985. International rules for seed testing. *Seed Sci. and Tech.* 13: 299-356.

Anonymous. 1990. Official methods of analysis of Association of official Analytical chemists. Vol. 2, 15th edition, 1990.

Anonymous. 2004, *Trigonella corniculata* Linn., Wealth of India, Raw materials, CSIR, New Delhi, 5 : 252-258.

Anonymous. 2015. Plant for a future. www.pfaf.org

Anupama G. 2012. Effect of nitrogen and spacing levels on growth and yield of kasuri methi (*Trigonella corniculata* L.) var. Pusa kasuri. M.Sc. (Horti) Thesis (Unpub.). Submitted to University of Horticultural Sciences, Bagalkot.

Anupama G., L. N. Hegde, N. K. Hegde, V. Devappa, A. B. Mastiholi and Sandhyarani Nishani. 2017. Effect of nitrogen and spacing levels on physiological and yield parameters of kasuri methi (*Trigonella corniculata* L.) var. Pusa kasuri. *Int. J. Curr. Microbiolog App. Sci.* **6**(9):723-733.

Babaleshwar S. B. and G. Raviraja Shetty. 2017. Influence of integrated nutrient management on nutrient uptake of kasuri methi (*Trigonella corniculata* L.), under hill zone of Bagalkot, Karnataka. *Int. J. Pure App. Biosci.* **5**(6): 1183-1188.

Bhardwaj K. K. 1996. Seed production studies in french bean (*Phaseolus vulgaris* L.) cultivar. M.Sc. (Agri.) Thesis (Unpub.). Submitted to Dr. Y. S. Parmar Uniersity of Horticulture and Forestry, Nauni, Solan (H. P.).

Bhutia P.H. and Sharangi A. B. 2016. Effect of dates of sowing and soil moisture level in different growth stages and yield dynamics of fenugreek (*Trigonella foenum-gracecum* L.). *Natl. Acad. Sci. Lett.* **39**(2):77–80.

Chopra S. L. and J. S. Kanwar. 1978. Analytical Agricultural Chemistry, Kalyani Publisher, Ludhiana, New Delhi. PP. 344.

Datta S. Kalam and R. Chatterjree. 2005. Effect of different levels of nitrogen and leaf cutting on growth, leaf and seed yield of fenugreek (*Trigonella corniculata* L.). *Indian J. of Agril. Scie.* **75**(9):580–581.

Datta Nilanjana, Jitesh K. Hore, Shreyasi Malik and Tapas Sarkar. 2017. Response of fenugreek (*Trigonella corniculata* L.) to different levels of nitrogen, phosphorus and potassium. *Curr. J. of App. Scie. and tech.* **22**(5):01–08.

Deora N. S., Jitendra Singh and M. L. Reager. 2009. Studies on nutrient management and seed rate on growth and herbage yield of fenugreek (*Trigonella corniculata* L.) cv. kasuri methi in Rajasthan. *J of spice and Aromatic Crops.* **18**(1):19–21.

Erum Shazia, Rshid Anwar and Shahid Masood. 2011. Evaluation of kasuri methi *trigonella foenum graecum* L. var. to establish GI right of Pakistans. *Pakistan J. Agric. Res.* **24**(4) PP: 25-30.

Gurjar J. S. 2005. Effect of various levels of NPK and FYM on yield and composition of kasuri methi (*Trigonella corniculata* L.) var. kasuri methi. M.Sc. Thesis (Unpub.) Jawaharlal Nehru Krishi Vishwa Vidyalaya, Jabalpur.

Godara A.S., S. Kapade and G. Laland Ravindra Singh. 2013. Effect of phosphorus and sulphur levels on the performance of Nagauri Methi (*Trigonella corniculata* L.) under semi-arid areas of Rajasthan. *Int. J. Seed Spices.* **3**(2) PP: 70-73.

*Gill B. S., G. S. Randhawa and S. S. Saini. 2001. Effect of sowing dates and herb cutting management on growth and yield of fenugreek (*Trigonella foenum-graecum* L.). *Indian J. Agron.* **46**(2): 364-367.

Jackson M. L. 1967. Chemical analysis. Prentice Hall of India Pvt. Ltd., New Delhi.

Jackson M. L. 1973. Soil chemical analysis. Prentice Hall of India Pvt. Ltd., New Delhi.

Jawarkar A. K. 2017. Effect of spacing and pinching on leaf yield of kasuri methi. M.Sc. (Horti.) Thesis (Unpub.). Submitted to Dr. Panjabarao Deshmukh Krishi Vidyapeeth, Akola (M. S.).

Kauser H., Bhoomika H. R. and Ibaad M. H. 2018. Interaction effect of different sowing dates and stage of pinching on growth, yield and economics of fenugreek. *Int. J. Pure App. Biosci.* **6**(2): 167-171.

*Korla B. N. and Amit Saini 2003. Effect of dates of sowing and cutting on seed yield of fenugreek. *Haryana J. Hort. Sci.* **32**(1/2):120-122.

Koyani C. R., P. K. Chowatia and B. S. Gohil. 2012. Effect of nitrogen and phosphorus on growth, yield attributes and yield of *rabi* fennel (*Foeniculum vulgare* Mill.). Agriculture: Towards a New Paradigm of Sustainability PP: 167-171.

Kumar A. and R. Singh. 2007. Response of fenugreek (*Trigonella foenum-graecum* L.) to different phosphorus and cutting management practices. *Indian J. of Agril. Sci.* **77**(3): 18-21.

Kumar R., S. S. Meena, R. K. Kakani, R. S. Mehta and N. K. Meena. 2015. Response of crop geometry, fertilizer levels and genotype on productivity of fenugreek (*Trigonella foenum-graecum* L.). *Int. J. of Seed Spices.* **5**(1): 63-67.

*Lal Sant, M. K. Rana and P. S. Pratap. 2003. Effect of dates of sowing and green cuttings on quality of different fenugreek (*Trigonella foenum-graecum* L.) genotypes. *Haryana J. Hort. Sci.* **32**(3/4):262-265.

Meena S.S., N. L. Sen and S. K. Malhotra. 2006. Influence of sowing date, nitrogen and plant growth regulators on growth and yield of coriander (*Coriandrum sutivum* L.). *J. of Spices and Aromatic Crops.* **15**(2): 88-92.

Meena N.K., S. S. Meena, Suman Gupta and G. Lal. 2018. Influence of different date of sowing, fertilizer level and weedicides on growth and yield of fenugreek (*Trigonella foenum-graecum* L.) under semi-arid condition. *Int. J.Curr. Microbiol. App. Sci.* **7**(8): 1844-1854.

Mitoo S. M. Masudur Rahman, Md. Hayder Khan Sujan and Faijul Islam. 2018. Effect of phosphorous and sulphur fertilization on seed yield of fenugreek (*Trigonella foenum*-graecum L.). *IOSR J. of Agril. Vet. Sci.* **11**(2) PP: 05-17.

Nandre. D. R. 2006. Effect of sowing dates and nutrient management on growth, seed yield and quality of fenugreek. Ph.D. (Hort) Thesis (Unpub.). Submitted to Dr. Panjabrao Deshmukh Krishi Vidyapeeth, Akola (M. S.).

Nandre. D. R. and R. G. Ghadge and B. S. Rajput. 2011. Effect of sowing dates and nutrient management on growth and seed yield of fenugreek. *Adv Res. J.Crop Improt.* **2**(2): 215-220.

Nandre. D. R. and V. S. Sahane. 2011. Effect of sowing dates and nutrient management on economics of seed production in fenugreek. *The Asian J. of Hort.* **6**(2): 459-461.

Olsen S. R., C. V. Cole, F. S. Watanabe and L. A. Dean. 1954. Estimation of available phosphorus in soil by extraction with sodium bicarbonate. *Circ. U.S. Dept. Agric.* pp. 939.

*Pan S., R. Chaterjee S. Datta, M. Bhattacharya, A. Pariari A. B. Sharangi and P. K. Chattopadhyay. 2002. Response of some cultivars of coriander (*Coriandrum sutivum* L.) to different dates of sowing. *South Indian Hort.* 51(1-6): 249-253.

Panse, V. G. and P. V. Sukhatme. 1985. Statistical methods for Agricultural workers. New Delhi, ICAR.

Pasricha V. and R. K. Gupta. (2014). Nutraceuticals potential of Methi *(Trigonella foenum graecum L.)* and kasuri methi (*Trigonella corniculata* L.).*J. of Pharmacognosy and Phytochemistry*. 3(4): 47-57.

Patrick B. and Yves S. 1996. Chemotaxonomic value of sterols and steroidal sapogenins in the genus Trigonella. *Biochemical systematic and Ecol*. **24**(2) PP: 157-164.

Piper C.S. 1966. Soil and Plant Analysis, Hans Pub. Bombay Asian Ed. PP 368-374.

Rana S. C., V. K. Pandita and Sanjai Sirohi. 2015. Influence of spacing and number of leaf cutting on seed yield in fenugreek. *Legume Res*. 38(6) PP 858:860.

Rongjie Z, Li W, Longxing W, Hongbin X and Shaoqing C. 2010. Determination of trigonelline in (*Trigonella foenum graecum* L.) by hydrophilic interaction chromatography. *Chinese. J Chromatography*. 28:379-382.

Seetharaman, S., B. C. Biswas, R. K. Tewatia and N. Prasad. 1994. Handbook of Fertilizer Usage. New Delhi, The fertilizer Association of India: 424-427.

Sharma S. K. 2006. Effect of different fertility levels on yield and nutrient uptake by kasuri methi (*Trigonella corniculata* L.). M.Sc. Thesis (Unpub.) Jawaharlal Nehru Krishi Vishwa Vidyalaya, Jabalpur.

Shoaf T. W. and Lium. B. W. 1976. Improved extraction of chlorophyll 'A' and 'B' from algae using dimethyl sulphoxide. Limnol. Oceanogr, 21: 926-926.

Sheoran R. S., H. C. Sharma, P. K. Panuu and Ram Niwas. 2000. Influence of sowing time and phosphorus on phenology, thermal recruitment and yield of fenugreek (*Trigonella foenum- graecum* L) genotype. *J. of Spice and Aromatic Crops*. **9**(1): 43-46.

Singh B., Kumar Nanavati, Sirpurkar W. M. and Nanavati V. S. 2017[a]. Standardization of date of sowing and stage of pinching on fenugreek yield and quality parameters. *Int. J. of Agril. Sci*. **7**(4):1309-1313.

Singh M. K., Ranjeet Kumar, Mukesh Kumar, Vipin Kumar and S. K. Lodhi. 2017[b]. Effect of nitrogen, phosphorus and cutting management on growth, flowering and seed yield of fenugreek (*Trigonella foenum-graecum* L.). *Annals of Horticultures*. **10**(2):196-201.

Singh S., M. R. Choudhari, O. P. Garhwal, M. L. Jakhar and B. L. Yadav. 2012. Effect of biofertilizers and inorganic sources of nitrogen and phosphorus on quality production of kasuri methi (*Trigonella corniculata* L.). *International J. Seed Spices*. **2**(2): 38-40.

Sowmya P. T. 2016. Effect of sowing dates and stage of pinching on growth, yield and quality of fenugreek (*Trigonella foenum-graecum* L.). M.Sc. Thesis (Unpub.) Rajmata Vijayaraje Scindia Krishi Vishwa Vidyalaya, Mandasuaur (Madhya Pradesh).

Subbaiah B.V. and G.L. Asija. 1956. A rapid procedure for the estimation of available nitrogen in soil. *Curr. Sci*. 25: 259.

Sunanda B. B., G. R. Shetty and J. Venkatesh. 2014. Influence of integrated nutrient management on growth, yield and quality of kasuri methi (*Trigonella corniculata* L.) under hill zone of Karnataka. *International J. Seed Spices* **4**(2) PP: 62-67.

Tandon H. L. S. 1993. Methods of analysis of soil plants, water and fertilizers. FAO and consultation organization, New Delhi.

*Varshney, I. P & Sood A. R. 1971. *J Indian*. Appl. Chem. Soc., 34, 208.

*Yadav J. S., Jagdev Singh, Virendra Kumar and B. D. Yadav. 2000. Effect of sowing time, spacing and seed rate on seed yield of fenugreek on light textured soil. *HAU J. Res.* **30**:107-111.

* Originals are not seen.

APPENDIX - I

Weekly Weather Data for the Year "2016-2017" recorded at Agromet Observatory, Dr. PDKV, Akola

MW	Dates	Tmax (°C)		Tmin (°C)		BSH (hrs)		WS (km/hr)		RHI (%)		RHII (%)		Evap (mm)		RF (mm)	
		N	A	N	A	N	A	N	A	N	A	N	A	N	A	N	A
40	1-7 Oct. 2016	33.9	29.1	20.2	22.8	8.1	4.6	7.5	1.8	81	92	45	73	5.5	3.2	21.8	61.5
41	8-14	34.1	31.2	18.7	21.3	4.2	7.6	4.1	3.0	76	90	40	59	5.3	4.1	16.0	29.0
42	15-21	33.9	32.9	18.1	16.5	8.4	8.9	4.4	0.4	74	80	36	29	5.5	4.4	3.1	0.0
43	22-28	33.1	32.5	18.5	15.8	8.4	8.4	4.1	0.9	73	80	36	34	5.3	4.7	10.0	0.0
44	29-4 Nov.	33.0	31.4	15.8	14.3	8.7	8.7	4.7	0.4	72	81	31	34	5.3	4.4	2.3	0.0
45	5-11	32.4	31.4	14.8	11.2	8.6	8.6	4.5	0.3	70	77	30	27	5.2	3.8	3.7	0.0
46	12-18	31.7	30.3	13.7	11.9	8.6	8.3	4.6	0.4	70	84	30	33	4.9	3.7	1.1	0.0
47	19-25	31.0	30.7	13.1	9.7	8.6	8.4	4.4	0.3	71	85	30	32	4.6	3.6	10.1	0.0
48	26-2 Dec.	30.3	31.9	12.4	10.9	8.8	8.9	4.6	0.2	71	85	31	31	4.3	3.8	6.8	0.0
49	3-9	29.8	30.1	11.2	10.7	8.7	8.0	4.7	0.8	70	88	29	35	4.3	3.5	1.3	0.0
50	10-16	29.4	30.2	10.3	10.7	8.8	8.0	4.5	1.2	70	79	27	32	4.2	4.2	1.3	0.0
51	17-23	29.5	29.2	10.6	8.6	8.7	8.5	4.7	0.5	69	85	29	35	4.3	3.5	0.9	0.0
52	24-31 Dec.16	29.2	29.5	10.7	8.4	8.6	8.5	4.8	0.5	70	83	31	29	4.3	3.4	2.6	0.0

1	1-7 Jan, 17	29.0	29.5	10.3	9.3	8.7	8.1	4.9	0.5	78	82	30	33	4.2	3.5	1.7	0.0
2	8-14	29.2	27.3	11.3	8.4	8.6	7.6	6.3	0.6	71	84	30	40	4.5	3.3	3.4	0.0
3	15-21	29.9	29.3	11.6	12.2	8.9	5.8	5.4	0.8	69	80	28	36	4.8	3.8	0.9	0.0
4	22-28	30.8	31.3	11.8	13.8	9.1	7.7	5.5	1.4	67	77	27	36	5.2	5.1	1.1	0.0
5	29-4 Feb.	31.1	31.6	12.1	11.8	9.3	8.8	5.8	0.8	61	80	25	27	5.6	4.8	2.8	0.0
6	5-11	31.3	32.5	11.9	14.2	9.1	8.5	5.6	1.7	59	72	23	27	5.9	5.7	4.9	0.0
7	12-18	32.5	33.2	13.4	16.4	9.4	7.7	6.1	2.2	56	74	22	27	6.6	6.2	0.1	0.5
8	19-25	33.0	35.1	13.8	14.8	9.5	9.4	6.5	2.9	57	59	22	14	7.3	8.3	3.3	0.0
9	26-4 March	34.7	36.4	14.8	15.6	9.6	9.2	7.0	1.4	50	42	17	12	8.1	8.0	3.4	0.0
10	5-11	36.1	34.7	16.7	17.4	9.6	9.0	6.8	4.2	44	46	18	16	9.0	9.5	2..1	0.0
11	12-18	37.3	34.2	17.5	16.0	9.6	8.3	6.9	3.5	42	42	17	14	9.5	9.1	2.5	0.0
12	19-25	38.5	37.7	18.3	18.9	9.6	9.1	6.9	2.7	37	35	13	13	10.5	10.0	0.3	0.0
13	26-1 April	39.0	42.5	19.7	22.5	9.6	9.0	7.6	1.8	36	32	15	11	11.3	10.6	2.9	0.0
14	2-8 Apr	40.0	40.7	21.7	26.2	9.4	9.1	7.3	6.4	37	33	14	15	11.7	13.6	1.0	0.0
15	9-15	40.8	41.9	23.1	20.0	9.5	9.5	8.4	2.2	35	27	14	8	12.9	11.3	0.4	0.0
16	16-22	41.6	43.9	24.1	27.7	9.7	10.0	8.6	7.1	36	25	14	6	13.9	17.5	0.5	0.0
17	23-29	42.3	41.1	25.4	25.5	9.8	10.1	9.0	8.2	37	35	15	10	14.7	14.9	0.5	0.0

N = Normal (1971-2000)

A = Actual (2005-06)

APPENDIX - II

Weakly Weather Data for the Year "2017-2018" recorded at Agromet Observatory, Dr. PDKV, Akola

MW	Dates	Tmax (°C)		Tmin (°C)		BSH (hrs)		WS (km/hr)		RHI (%)		RHII (%)		Evap (mm)		RF (mm)	
		N	A	N	A	N	A	N	A	N	A	N	A	N	A	N	A
40	1-7 Oct 2017	33.7	35.6	21.2	21.1	7.6	8.0	4.8	1.2	82	81	47	42	5.4	6.5	23.4	0.0
41	8-14	34.0	31.9	19.8	23.2	8.1	4.3	4.5	0.6	78	91	40	60	5.3	4.1	13.1	57.5
42	15-21	33.7	33.7	18.3	20.6	8.2	6.9	4.6	0.8	76	71	37	43	5.3	5.6	6.1	0.0
43	22-28	33.1	33.9	16.8	16.6	8.3	8.5	4.4	0.6	74	76	34	29	5.3	5.8	7.6	4.5
44	29-4 Nov	32.7	32.7	16.0	14.0	8.4	8.5	4.1	0.6	73	82	32	26	5.3	5.9	2.3	0.0
45	5-11	32.3	31.6	15.2	14.7	8.4	8.5	3.9	1.0	71	81	32	28	5.1	5.4	3.0	0.0
46	12-18	31.6	31.0	14.6	13.7	8.3	7.9	3.9	0.4	73	79	32	37	4.8	4.9	5.3	0.0
47	19-25	31.0	31.8	13.3	17.9	8.4	5.3	3.7	0.9	72	84	30	40	4.6	4.6	7.7	0.0
48	26-2 Dec	30.5	30.6	12.8	11.2	8.4	8.2	3.6	0.9	71	79	32	19	4.4	5.4	5.5	0.0
49	3-9	30.0	29.7	11.9	14.5	8.4	5.6	3.8	1.5	71	78	30	38	4.3	4.8	1.0	0.0
50	10-16	29.6	31.2	10.9	13.7	8.4	7.6	3.6	0.8	71	74	28	27	4.2	5.2	0.8	0.0
51	17-23	29.5	29.8	10.8	10.4	8.5	6.6	3.8	0.8	70	80	29	25	4.1	4.6	0.9	0.0
52	24-31	29.1	29.4	11.1	8.7	8.3	8.2	4.5	1.0	71	72	30	22	4.2	4.3	2.6	0.0
1	1-7 Jan 18	28.8	29.5	11.0	9.4	8.2	7.5	4.4	0.8	71	74	31	23	4.2	3.8	2.8	0.0

2	8-14	29.3	29.6	11.7	12.3	8.3	7.7	4.4	1.2	71	58	30	23	4.4	5.9	3.3	0.0
3	15-21	30.0	31.3	12.0	12.1	8.6	7.9	4.5	1.3	68	67	28	20	4.9	6.4	0.7	0.0
4	22-28	30.6	29.8	12.0	10.0	8.8	8.3	4.6	1.7	65	57	26	20	5.2	7.3	0.9	0.0
5	29-4 Feb	31.0	32.4	12.6	10.9	8.8	6.9	4.9	1.0	62	59	25	15	5.5	6.7	3.0	0.0
6	5-11	31.4	32.3	12.7	16.0	8.8	3.8	5.0	1.9	59	47	23	20	5.9	7.4	3.7	0.0
7	12-18	32.7	29.9	14.4	14.9	9.0	6.1	5.4	3.9	55	70	22	33	6.6	6.6	0.1	0.7
8	19-25	33.4	35.5	14.5	17.9	9.1	8.0	5.7	1.8	54	48	21	19	7.3	7.4	2.5	0.0
9	26-4 Mar	35.0	37.3	15.7	18.5	9.5	8.1	6.1	1.7	50	40	18	15	8.2	7.6	4.1	0.0
10	5-11	35.9	36.7	17.3	22.1	9.2	6.1	6.1	4.0	46	37	20	20	8.8	8.9	5.2	3.0
11	12-18	37.0	34.7	18.1	20.3	9.1	4.8	6.3	2.5	45	56	18	18	9.2	7.3	2.4	0.4
12	19-25	38.4	37.7	19.3	20.4	9.2	8.6	6.4	3.3	39	36	15	14	10.4	10.2	0.6	0.0
13	26-1 Apr	39.0	41.1	20.4	20.0	9.2	9.2	6.9	3.2	37	27	15	7	11.2	11.1	2.2	0.0
14	2-8 Apr	40.0	40.9	21.7	22.7	9.4	7.7	7.3	2.4	37	27	14	11	11.7	10.1	1.0	0.0
15	9-15	40.8	40.4	23.1	24.5	9.5	8.6	8.4	2.9	35	45	14	16	12.9	10.4	0.4	0.3
16	16-22	41.6	42.3	24.1	27.8	9.7	9.9	8.6	5.6	36	30	14	12	13.9	13.8	0.5	0.0
17	23-29	42.3	42.9	25.4	26.4	9.8	10.1	9.0	4.2	37	28	15	7	14.7	13.1	0.5	0.0

N = Normal (1971-2000)

A = Actual (2006-07)

APPENDIX - III

Interaction effects of dates of sowing and fertilizer doses on days to 50 per cent flowering in kasuri methi

Treatments	Days to 50 per cent flowering		
	2016-17	2017-18	Pooled
D_1F_1	59.00	58.01	58.51
D_1F_2	58.26	57.40	57.83
D_1F_3	58.78	57.56	58.17
D_1F_4	58.16	57.08	57.62
D_2F_1	60.75	59.47	60.11
D_2F_2	59.63	58.28	58.96
D_2F_3	60.41	58.91	59.66
D_2F_4	59.55	58.35	58.95
D_3F_1	57.98	56.82	57.40
D_3F_2	57.33	56.12	56.73
D_3F_3	57.62	56.36	56.99
D_3F_4	57.05	55.64	56.34
D_4F_1	56.82	55.41	56.12
D_4F_2	56.13	55.02	55.58
D_4F_3	56.54	55.23	55.89
D_4F_4	55.73	54.81	55.27
D_5F_1	54.78	54.66	54.72
D_5F_2	54.51	54.19	54.35
D_5F_3	54.68	54.49	54.58
D_5F_4	54.34	54.06	54.20
D_6F_1	54.77	54.26	54.51
D_6F_2	54.26	53.70	53.98
D_6F_3	54.25	53.95	54.10
D_6F_4	54.12	53.44	53.78
'F' test	NS	NS	NS
SE(m) ±	0.18	0.41	0.23
CD at 5%	-	-	-

APPENDIX - IV

Interaction effects of dates of sowing and fertilizer doses on number of seeds per pod in kasuri methi

Treatments	Number of seeds per pod		
	2016-17	2017-18	Pooled
D_1F_1	5.43	5.21	5.32
D_1F_2	5.92	5.64	5.78
D_1F_3	5.89	5.34	5.62
D_1F_4	5.94	5.70	5.82
D_2F_1	5.22	5.10	5.16
D_2F_2	5.88	5.31	5.60
D_2F_3	5.35	5.16	5.26
D_2F_4	5.89	5.35	5.62
D_3F_1	4.24	4.81	4.53
D_3F_2	4.71	4.77	4.74
D_3F_3	4.43	4.71	4.57
D_3F_4	4.79	4.94	4.86
D_4F_1	3.34	3.99	3.67
D_4F_2	3.90	4.24	4.07
D_4F_3	3.93	4.37	4.15
D_4F_4	4.08	4.47	4.27
D_5F_1	2.33	2.80	2.57
D_5F_2	3.03	3.62	3.33
D_5F_3	3.05	3.16	3.10
D_5F_4	3.24	3.96	3.60
D_6F_1	2.56	3.33	2.94
D_6F_2	2.25	3.01	2.63
D_6F_3	2.51	3.79	3.15
D_6F_4	2.62	3.29	2.96
'F' test	Sig.	NS	NS
SE(m) ±	0.13	0.18	0.12
CD at 5%	0.38	-	-

APPENDIX - V

Interaction effects of dates of sowing and fertilizer doses on test weight of seed (g) in kasuri methi

Treatments	Test weight of seed (g)		
	2016-17	2017-18	Pooled
D_1F_1	1.172	1.176	1.174
D_1F_2	1.176	1.178	1.177
D_1F_3	1.175	1.177	1.176
D_1F_4	1.176	1.179	1.178
D_2F_1	1.173	1.176	1.174
D_2F_2	1.175	1.177	1.176
D_2F_3	1.173	1.176	1.175
D_2F_4	1.175	1.177	1.176
D_3F_1	1.171	1.173	1.172
D_3F_2	1.173	1.175	1.174
D_3F_3	1.172	1.174	1.173
D_3F_4	1.173	1.175	1.174
D_4F_1	1.168	1.170	1.169
D_4F_2	1.169	1.170	1.170
D_4F_3	1.171	1.171	1.171
D_4F_4	1.171	1.172	1.171
D_5F_1	1.162	1.167	1.165
D_5F_2	1.167	1.168	1.168
D_5F_3	1.167	1.169	1.168
D_5F_4	1.168	1.170	1.169
D_6F_1	1.164	1.165	1.164
D_6F_2	1.165	1.166	1.166
D_6F_3	1.165	1.166	1.166
D_6F_4	1.166	1.167	1.167
'F' test	NS	NS	NS
SE(m) ±	0.0010	0.0011	0.0008
CD at 5%	-	-	-

APPENDIX - VI

Interaction effects of dates of sowing and fertilizer doses on total chlorophyll content in kasuri methi (mg/100 mg of leaves at 60 DAS)

Treatments	Chlorophyll content (mg/100mg of leaves)		
	2016-17	2017-18	Pooled
D_1F_1	2.93	2.99	2.96
D_1F_2	3.01	3.12	3.07
D_1F_3	3.05	3.18	3.12
D_1F_4	3.10	3.19	3.14
D_2F_1	2.82	2.93	2.88
D_2F_2	2.70	2.98	2.84
D_2F_3	2.83	3.16	3.00
D_2F_4	2.96	3.16	3.06
D_3F_1	2.56	2.74	2.65
D_3F_2	2.58	2.84	2.71
D_3F_3	2.64	2.87	2.75
D_3F_4	2.68	2.83	2.75
D_4F_1	2.72	2.82	2.77
D_4F_2	2.75	2.92	2.84
D_4F_3	2.87	2.79	2.83
D_4F_4	2.83	2.90	2.87
D_5F_1	2.69	2.74	2.72
D_5F_2	2.73	2.80	2.77
D_5F_3	2.79	2.51	2.65
D_5F_4	2.74	2.84	2.79
D_6F_1	2.68	2.71	2.69
D_6F_2	2.67	2.73	2.70
D_6F_3	2.65	2.75	2.70
D_6F_4	2.67	2.73	2.70
'F' test	NS	NS	NS
SE(m) ±	0.12	0.07	0.12
CD at 5%	-	-	-

APPENDIX - VII

Interactions effect of dates of sowing and fertilizer dose on total dry matter (%) in kasuri methi

Treatments	Total dry matter (%)		
	2016-17	2017-18	Pooled
D_1F_1	14.12	15.27	14.69
D_1F_2	15.49	15.82	15.66
D_1F_3	14.96	15.62	15.29
D_1F_4	15.62	15.97	15.79
D_2F_1	13.70	14.72	14.21
D_2F_2	14.30	15.39	14.84
D_2F_3	13.91	15.07	14.49
D_2F_4	14.51	15.78	15.15
D_3F_1	12.84	13.89	13.37
D_3F_2	13.23	14.17	13.70
D_3F_3	13.03	13.99	13.51
D_3F_4	13.45	14.34	13.90
D_4F_1	11.29	12.99	12.14
D_4F_2	12.12	13.54	12.83
D_4F_3	11.58	13.20	12.39
D_4F_4	12.34	13.72	13.03
D_5F_1	10.51	12.35	11.43
D_5F_2	10.87	12.74	11.81
D_5F_3	10.72	12.58	11.65
D_5F_4	11.10	12.87	11.98
D_6F_1	9.14	10.76	9.95
D_6F_2	10.09	11.79	10.94
D_6F_3	9.56	11.70	10.63
D_6F_4	10.32	12.16	11.24
'F' test	NS	NS	NS
SE(m) ±	0.54	0.35	0.25
CD at 5%	-	-	-

APPENDIX - VIII

Cost of cultivation of kasuri methi seed production per hectare

Sr. No.	Cost of expenditure	Male	Female	Tractor	Total (Rs ha^{-1})
1.	Land Preparation				
a)	Ploughing	1		Rs. 800 (6 hr)	4800
b)	Harrowing	1		Rs. 800 (5 hr)	4000
c)	Flat bed	1		Rs. 500 (8 hr)	4000
2.	Manuring				
a)	Cost of manure				15000
b)	Spreading of manure	8			1840
3.	Seed material				
a)	Cost of seed				600
b)	Cost of seed sowing		10		2300
4.	Irrigation	32			7820
5.	Weeding		60		13800
6.	Harvesting		30		6900
7.	Threshing		20		4600
8.	Land revenue				
9.	Miscellaneous charges				7442
	Total cost				73102

Note : The cost of nutrients management treatment will be according to the treatments

Cost of nutrients :

1. Urea - Rs. 6 kg^{-1}
2. Single super phosphate - Rs. 7.4 kg^{-1}
3. Muriate of potash - Rs. 12.6 kg^{-1}

Labour charges - Rs. 230 day^{-1}

Selling price of kasuri methi seeds - Rs. 500 kg^{-1}

- Rs. 50,000 q^{-1}

INDEX